T0181456

Springer Tracts in Nature-Inspired Computing

Series Editors

Xin-She Yang, School of Science and Technology, Middlesex University, London, UK

Nilanjan Dey, Department of Information Technology, Techno India College of Technology, Kolkata, India

Simon Fong, Faculty of Science and Technology, University of Macau, Macau, Macao

The book series is aimed at providing an exchange platform for researchers to summarize the latest research and developments related to nature-inspired computing in the most general sense. It includes analysis of nature-inspired algorithms and techniques, inspiration from natural and biological systems, computational mechanisms and models that imitate them in various fields, and the applications to solve real-world problems in different disciplines. The book series addresses the most recent innovations and developments in nature-inspired computation, algorithms, models and methods, implementation, tools, architectures, frameworks, structures, applications associated with bio-inspired methodologies and other relevant areas.

The book series covers the topics and fields of Nature-Inspired Computing, Bio-inspired Methods, Swarm Intelligence, Computational Intelligence, Evolutionary Computation, Nature-Inspired Algorithms, Neural Computing, Data Mining, Artificial Intelligence, Machine Learning, Theoretical Foundations and Analysis, and Multi-Agent Systems. In addition, case studies, implementation of methods and algorithms as well as applications in a diverse range of areas such as Bioinformatics, Big Data, Computer Science, Signal and Image Processing, Computer Vision, Biomedical and Health Science, Business Planning, Vehicle Routing and others are also an important part of this book series.

The series publishes monographs, edited volumes and selected proceedings.

More information about this series at http://www.springer.com/series/16134

Aziz Ouaarab

Discrete Cuckoo Search
for Combinatorial
Optimization

 Springer

Aziz Ouaarab
Ecole Supérieure de Technologie
d'Essaouira
Essaouira, Morocco

ISSN 2524-552X   ISSN 2524-5538 (electronic)
Springer Tracts in Nature-Inspired Computing
ISBN 978-981-15-3838-4   ISBN 978-981-15-3836-0 (eBook)
https://doi.org/10.1007/978-981-15-3836-0

This Springer imprint is published by the registered company Springer Nature Singapore Pte Ltd.
The registered company address is: 152 Beach Road, #21-01/04 Gateway East, Singapore 189721,
Singapore

Preface

Since the first appearance, discrete cuckoo search for the traveling salesman problem has attracted attention of several researchers, and Ph.D. and graduate students. I have received many emails to explain how DCS works and to share the computer code of DCS, and respectively, all the contributions presented in this book. For this reason, I have then reported the description with examples of implementation and computer code on how CS is adapted to solve different combinatorial optimization problems.

The book starts by the theoretical sides in order to present the main ideas that guide the conception of the approach from how it presents the solution and how it changes it, to applying advanced strategies such as intensification/diversification, Lévy flights' displacement, and population smart cuckoo search process. Adapting CS to a set of problems is also presented with sufficient details.

This book has as the main objective facilitating the reuse of CS to be applied in different combinatorial optimization problems. It represents a useful tool for both researchers and students to solve other problems with less adaptation constraints. In addition, its simplified computer code version can be easily improved and designed to be reused.

Marrakech, Morocco
February 2020

Aziz Ouaarab

Acknowledgements

I am grateful to my mentors and co-authors Xin-She Yang and Belaïd Ahiod for their help, advice, and guidance on the contributions reported in this book.

I would like to thank my Editor, Aninda Bose, and Project Coordinator, Ramamoorthy Rajangam, for their help and professionalism.

None of this would have been possible without the help of my family. I would like to thank my parents, sisters, and brothers, whose love and assistance are with me in whatever I pursue.

Last but not least, I thank my loving and supportive wife.

Aziz Ouaarab

Acknowledgements

Contents

About the Author

Dr. Aziz Ouaarab is an Assistant Professor of Computer Sciences and Applied Mathematics at Cadi Ayyad University of Marrakech, Morocco. Dr. Ouaarab received his M.S. degree in 2011 and his D.Phil. in Engineering Sciences in 2015, both from Mohammed V University of Rabat, Morocco. His thesis project focused on the use of nature-inspired metaheuristics to solve combinatorial optimization problems, and on cuckoo search as a special case to be improved. He joined the Computer Engineering and Systems Laboratory at the Faculty of Sciences and Techniques, Marrakech, in 2019. His research chiefly focuses on combinatorial optimization problems and how they can be solved by means of swarm intelligence and nature-inspired metaheuristics. Dr. Ouaarab currently serves on the editorial board of the International Journal of Bio-Inspired Computation.

Acronyms

ABC	Artificial Bee Colony
ACO	Ant Colony Optimization
BCO	Bee Colony Optimization
Bkv	Best known value
COP	Combinatorial Optimization Problem
CS	Cuckoo Search
CX	Crossover
DCS	Discrete Cuckoo Search
DG	Distributed Generation
FA	Firefly Algorithm
GA	Genetic Algorithms
gBest	global Best
I/D	Intensification/Diversification
JSSP	Job Shop Scheduling Problem
lBest	local Best
OR-Library	Operations Research Library
pBest	particle Best
PSO	Particle Swarm Optimization
QAP	Quadratic Assignment Problem
QAPLIB	Quadratic Assignment Problem Library
RK	Random Key
RKCS	Random-Key Cuckoo Search
TSP	Traveling Salesman Problem
TSPLIB	Traveling Salesman Problem Library

Chapter 1
Introduction

1.1 General Introduction

The need to optimize, plan, or make decisions in real time is everywhere, even in our daily lives. At all moments and situations, we are obliged to make a decision among many options. The problem is that sometimes our decision depends on a multitude of parameters and constraints, which makes the verification of all possible choices more difficult. Replacing the decision-making context of our daily lives by that of large companies and mega-industries, makes gains, and losses increase proportionally. Dealing with these optimization problems is done by using a variety of methods that perform different tools. These methods can be classified by considering two important performance measures: computation time and solution quality. Methods which consume less computation time are relatively more forgiving with the optimality regarding the solution quality. On the other side, methods requiring high-quality solutions are qualified as greedy, in terms of calculation time. The optimality and the calculation time are mainly linked to the performance of the method and the robustness of existing material resources. This situation requires an optimization of the problem conception and the behavior of methods vis-à-vis limitations caused by the hardware.

The nature of the optimization problems can be associated with that of their search/solution space. In general, there are two types of search spaces: continuous and discrete (or combinatorial) space. The first is the space containing solutions characterized by their continuous or real variables, while the second is a set of solutions of discrete variables. A large part of problems in planning, logistics, transport, finance, and manufacturing, which are of discrete or combinatorial nature, represent a real challenge of calculation for algorithms developed to solve them, knowing that the size of these problems exceeds, more and more with industry development. Indeed, several optimization problems are apparently simple, but their difficulty increases exponentially with their size. This makes the choice to search and verify each combination (solution) not practical.

© Springer Nature Singapore Pte Ltd. 2020
A. Ouaarab, *Discrete Cuckoo Search for Combinatorial Optimization*,
Springer Tracts in Nature-Inspired Computing,
https://doi.org/10.1007/978-981-15-3836-0_1

Talking complexity, NP-hard problems (Arora and Barak 2009) belong to the class of problems that attracts researchers in the field of optimization to propose new efficient approaches. However, no algorithm is effective in dealing with this type of problems. The need to quickly find a good quality solution favors the appearance of approximate or stochastic algorithms. In this context, metaheuristics (Blum and Roli 2003; Glover and Kochenberger 2003) have shown their high performance for a wide variety of optimization problems and have more advantages compared to traditional algorithms, such as the ability to deal with high levels of complexity, and to be adapted (with a set of constraints) to several problem models and their application in several real-world areas, starting from operational research, passing through engineering and artificial intelligence (Yang and Gandomi 2012; Yang et al. 2013; Gandomi et al. 2011, 2012), where there is a need for optimize digital functions, and systems containing a large number of parameters to be tuned simultaneously.

1.2 Challenges in Metaheuristics

Metaheuristics are a family of algorithms that carries out a search strategy which combines an exploration of the search space on its global scale, with an intensified exploitation in regions considered promising. They employ one or a population of interactive agents for iterative search, of new potentially better or optimal solutions. The majority of recently designed metaheuristics are inspired by nature (Yang and Deb 2009; Yang 2009), imitating biological phenomena (Wang et al. 2000), physical (Van Laarhoven and Aarts 1987), or social (Rao et al. 2011). Among the strong points of metaheuristics, we cite their relative ease of implementation, practically adaptable to several types of problems while producing solutions of good quality or even of optimal quality in a reasonable time. While searching scholastically toward a global optimum, they scan the search space (of solutions) by moving from one solution to another in the hope of finding better. In order to be more productive in their movement, metaheuristics need to learn about the spatial topology of the problem, starting with the notion of neighborhood. Two so-called neighboring solutions in the search space, when they contain coordinates of relatively close values and their quality values, are also close to each other. This condition is achievable in the majority of continuous optimization problems. However, in the case of combinatorial optimization, the debate remains open. The displacement, in continuous space, of the current solution toward a neighboring solution is performed by a change in its coordinates. This displacement generally produces a small difference between the two qualities of the current solution and its neighbor. On the other hand, a small change on a combinatorial solution space can lead to a "neighboring" solution of remarkably different qualitative values. This explains the possibility of finding the optimal solution and a poor quality solution in the same neighborhood. Most of metaheuristics are developed mainly to solve continuous optimization problems. All these metaheuristics that deal with combinatorial optimization problems are attempting to be adapted to this category of problems. To go directly to the heart of the problem of adaptation to combinatorial spaces, we say that a successful adaptation is the one that shows the highest consistency between two neighboring solutions and their fitness values. In fact, while moving from a current solution to its neighbor, the notion of step length requires a clear definition about the difference between a small and a large step. All these conditions mentioned, in order to produce an easy and efficient adaptation, are not fully validated in most cases of metaheuristics adaptations (designed mainly for continuous space) to combinatorial spaces. And in the rest of the cases, these notions are vague and undetectable. This makes understanding how to converge toward optimal solutions, a very difficult task which slows down the improvement of metaheuristics adapted to this type of space. To deal with the issue of adaptation, three different approaches are proposed. The first is to design metaheuristics that solve naturally combinatorial problems and do not need any adaptation to this space such as ant systems (Stützle and Hoos 2000). In this approach, a combinatorial solution is constructed with respect to the treated problem and do not need further adaptations. The second approaches are initially designed and validated by continuous problems and then performed a discretization process to search in the combinatorial space. These algorithms represent the majority in the literature by a variety of techniques and adaptation strategies. We can cite cuckoo search discretization (Ouaarab et al. 2014a; Ouyang et al. 2013; Gherboudj et al. 2012; Ouaarab et al. 2015a), firefly algorithm (Jati et al. 2011; Marichelvam et al. 2013; Osaba et al. 2017; Sayadi et al. 2013), particle swarm optimization (Zhong et al. 2007; Gomez et al. 2006; Lin et al. 2010; Huilian 2010), and bat algorithm (Osaba et al. 2016; Luo et al. 2014; Zhou et al. 2018). They are mainly based on how to search in the combinatorial space instead of constructing the solution (they use primitive methods to generate quickly the initial solution). So, adaptations will be performed on how to move in the combinatorial space and not how to construct a solution. The third category of adaptation keeps searching in the continuous space and the found solutions are projected in a parallel combinatorial space by using an encoding scheme technique such as Debels et al. (2006), Chen et al. (2011), and Ouaarab et al. (2015b).

The challenge in these three categories is reaching the perfect adaptation for the treated problem while still independent vis-à-vis its constraints. The adaptation has to be general for the majority of combinatorial problems and efficient for each solved problem.

1.3 Book Overview

The idea of joining the algorithms improvements for finding better solutions, to a relevant conception of the search space topology, is the reason for discussing contributions presented in this book. Indeed, the main notions developed are related on how a metaheuristic, initially designed for continuous spaces, behaves in the combinatorial one. Considering this notion, the following concepts lack a clear and meaningful definition in several of literature-developed metaheuristics: solution, step, and neighborhood.

Solution: It represents the pillar of each optimization procedure. The proposed approach gives a definition of solution and its representation so as to show the impact of simplifying its model vis-à-vis the studied problem and the calculation of its quality.

Step: It is simply the unit of distance between two solutions in the search space. It is useful for defining neighborhood notion.

Neighborhood: It is the set of close solutions to a given solution in the search space. These neighbor solutions are reached by a movement with the smallest step.

The first set of contributions developed in this book adopts a strategy of reduce dependences between problems and algorithm by considering the treated problem as a black box. The input of the box is fed by steps or displacements, and the output is the quality of the found solution. The realization of these contributions takes as reference, cuckoo search algorithm (Yang and Deb 2009) via Lévy flights. Indeed, before the conception of the discrete version of cuckoo search, an improvement is proposed and validated by a set of applications on three typical combinatorial optimization problems. The objective of these applications is to show the simplicity of proceeding without a detailed consideration of the treated problem. All the contributions of this approach can be presented as follows:

Improved cuckoo search (Ouaarab et al. 2014a): Inspired by the behavior of some more intelligent cuckoos, the proposed improvement is a restructuring of the population. The aim of this improvement is, at the same time, the reinforcement of intensified research around promising regions and of independence from the best solution of the population to better explore the research space.

Discrete cuckoo search (Ouaarab et al. 2014a): The central point of this contribution is the passage from continuous space to that of combinations. It aims to project the policy of a metaheuristic (designed for a continuous space) concerning the notions of intensification/diversification through its displacement and its conception of the neighborhood, on a given combinatorial optimization problem.

Application on traveling salesman problem (Ouaarab et al. 2014b): The traveling salesman problem represents a typical model in operational research for testing the performance of an optimization algorithm. The objective of this application is to confirm the theoretical part of the design, followed by a validation reported in the digital test results.

Application on job shop scheduling problem (Ouaarab et al. 2014c, 2015a): This application is the gate to a multitude of other scheduling problems. It supports our theory on the ease of adapting discrete cuckoo search to the majority of combinatorial optimization problems. The experimental results of this approach are also very encouraging.

Application on quadratic assignment problem (Ouaarab et al. 2015c): This fundamental combinatorial optimization problem which represents a generalization of several problems is a model for others from different real fields of application. Validation test results of the adapted discrete cuckoo search are illustrated.

The second group of reported contributions represents a proposal for another adaptation category. It adds to cuckoo search a random-key encoding scheme to project its movements on the combinatorial space. A description of the main strategies of this approach is provided with a couple of application in the following order:

Random-key cuckoo search (Ouaarab et al. 2015b): The idea is to project the real values (continuous space) generated by Lévy flights, directly into the combinatorial space. In this approach, the balance control between intensification and diversification is done in the continuous space, and then the found solutions are projected in the combinatorial space.

Application on traveling salesman and quadratic assignment problems (Ouaarab et al. 2015b, d): Random-key cuckoo search is applied on traveling salesman and quadratic assignment problems. The quality of the test results shows encouraging performances for both problems.

1.4 Book Organization

This book is divided into two parts. The first shows theoretical sides of all the strategies implemented in the second part. It introduces, in the first chapter, combinatorial optimization space and provides examples of combinatorial problems: (1) traveling salesman problem, (2) job shop scheduling problem, and (3) quadratic assignment problem.

The second chapter discusses how combinatorial problems are solved. Two groups of resolution approaches are presented: Vertical and horizontal and three adaptation strategies: Discrete, discreted, and projected.

The third chapter describes how cuckoo search is adapted to combinatorial space. It presents a complete approach starting from the natural phenomenon toward an algorithm adaptation model. The chapter begins by dissecting the source of inspiration of cuckoo search to isolate the components to be adapted for the combinatorial space.

The second part is composed of two validation chapters. The first chapter validates discrete cuckoo search over an application on traveling salesman problem, job shop scheduling problem, and the quadratic assignment problem. In this chapter, the same logic is followed while adapting the algorithm for all the three problems. The steps of matching the components of the problem and those of the algorithm while keeping a high level of performance are shown.

The second chapter deals with another level of adaptation introduced in random-key cuckoo search. The aim of this approach is to directly represent the values generated by Lévy flights, in the combinatorial space. A detailed description of this proposition is reported followed by an application on traveling salesman problem and quadratic assignment problem before numerical result validations.

References

Arora S, Barak B (2009) Computational complexity: a modern approach. Cambridge University Press

Blum C, Roli A (2003) Metaheuristics in combinatorial optimization: overview and conceptual comparison. ACM Comput Surv (CSUR) 35(3):268–308

Chen H, Li S, Tang Z (2011) Hybrid gravitational search algorithm with random-key encoding scheme combined with simulated annealing. Int J Comput Sci Netw Secur 11:208–217

Debels D, De Reyck B, Leus R, Vanhoucke M (2006) A hybrid scatter search/electromagnetism meta-heuristic for project scheduling. Eur J Oper Res 169(2):638–653

Gandomi AH, Talatahari S, Yang X-S, Deb S (2012) Design optimization of truss structures using cuckoo search algorithm. In: The structural design of tall and special buildings

Gandomi AH, Yang X-S, Alavi AH (2011) Mixed variable structural optimization using firefly algorithm. Comput Struct 89(23–24):2325–2336

Gherboudj A, Layeb A, Chikhi S (2012) Solving 0–1 knapsack problems by a discrete binary version of cuckoo search algorithm. Int J Bio-Inspired Comput 4(4):229–236

Glover F, Kochenberger GA (2003) Handbook of metaheuristics. Springer

Gomez A, Gonzalez R, Parreo J, Pino R (2006) A particle swarm-based metaheuristic to solve the travelling salesman problem. In: International conference on artificial intelligence, pp 698–702

Huilian F (2010) Discrete particle swarm optimization for TSP based on neighborhood. J Comput Inf Syst 6(10):3407–3414

Jati GK et al (2011) Evolutionary discrete firefly algorithm for travelling salesman problem. Springer

Lin T-L, Horng S-J, Kao T-W, Chen Y-H, Run R-S, Chen R-J, Lai J-L, Kuo I-H (2010) An efficient job-shop scheduling algorithm based on particle swarm optimization. Expert Syst Appl 37(3):2629–2636

Luo Q, Zhou Y, Xie J, Ma M, Li L (2014) Discrete bat algorithm for optimal problem of permutation flow shop scheduling. Sci World J

Marichelvam MK, Prabaharan T, Yang XS (2013) A discrete firefly algorithm for the multi-objective hybrid flowshop scheduling problems. IEEE Trans Evol Comput 18(2):301–305

Osaba E, Yang X-S, Diaz F, Lopez-Garcia P, Carballedo R (2016) An improved discrete bat algorithm for symmetric and asymmetric traveling salesman problems. Eng Appl Artif Intell 48:59–71

Osaba E, Yang X-S, Diaz F, Onieva E, Masegosa AD, Perallos A (2017) A discrete firefly algorithm to solve a rich vehicle routing problem modelling a newspaper distribution system with recycling policy. Soft Comput 21(18):5295–5308

Ouaarab A, Ahiod B, Yang X-S (2014a) Discrete cuckoo search algorithm for the travelling salesman problem. Neural Comput Appl 24(7–8):1659–1669

Ouaarab A, Ahiod B, Yang X-S (2014b) Improved and discrete cuckoo search for solving the travelling salesman problem. In: Yang X-S (ed) Cuckoo search and firefly algorithm. Studies in computational intelligence, vol 516. Springer International Publishing, pp 63–84

Ouaarab A, Ahiod B, Yang X-S, Abbad M (2014c) Discrete cuckoo search algorithm for job shop scheduling problem. In: 2014 IEEE international symposium on intelligent control (ISIC), pp 1872–1876

Ouaarab A, Ahiod B, Yang X-S (2015a) Discrete cuckoo search applied to job shop scheduling problem. In: Yang X-S (ed) Recent advances in swarm intelligence and evolutionary computation. Studies in computational intelligence, vol 585. Springer International Publishing, pp 121–137

Ouaarab A, Ahiod B, Yang X-S (2015b) Random-key cuckoo search for the travelling salesman problem. Soft Comput 19(4):1099–1106

Ouaarab A, Ahiod B, Yang X-S, Abbad M (2015c) Discrete cuckoo search for the quadratic assignment problem. In: 11th metaheuristics international conference

Ouaarab A, Ahiod B, Yang X-S, Abbad M (2015d) Random-key cuckoo search for the quadratic assignment problem (submitted). Nat Comput

Ouyang X, Zhou Y, Luo Q, Chen H (2013) A novel discrete cuckoo search algorithm for spherical traveling salesman problem. Appl Math Inf Sci 7(2)

Rao RV, Savsani VJ, Vakharia D (2011) Teaching-learning-based optimization: a novel method for constrained mechanical design optimization problems. Comput Aided Des 43(3):303–315

Sayadi MK, Hafezalkotob A, Naini SGJ (2013) Firefly-inspired algorithm for discrete optimization problems: an application to manufacturing cell formation. J Manuf Syst 32(1):78–84

Stützle T, Hoos HH (2000) Max-min ant system. Futur Gener Comput Syst 16(8):889–914

Van Laarhoven PJ, Aarts EH (1987) Simulated annealing. Springer

Wang L, Pan J, Jiao L (2000) The immune algorithm. Acta Electron Sin 28(7):74–78

Yang X-S (2009) Firefly algorithm, lvy flights and global optimization, pp 209–218

Yang X-S, Cui Z, Xiao R, Gandomi AH, Karamanoglu M (2013) Swarm intelligence and bio-inspired computation: theory and applications. Newnes

Yang XS, Deb S (2009) Cuckoo search via lévy flights. In: World congress on nature and biologically inspired computing, 2009. NaBIC 2009. IEEE, pp 210–214

Yang X-S, Gandomi AH (2012) Bat algorithm: a novel approach for global engineering optimization. Eng Comput 29(5):464–483

Zhong W, Zhang J, Chen W (2007) A novel discrete particle swarm optimization to solve traveling salesman problem. In: IEEE congress on evolutionary computation, 2007. CEC 2007. IEEE, pp 3283–3287

Zhou X, Zhao X, Liu Y (2018) A multiobjective discrete bat algorithm for community detection in dynamic networks. Appl Intell 48(9):3081–3093

Part I
Theory and Formulations

Chapter 2
Combinatorial Optimization Space

For an optimization problem, the search process for the optimal solution is generally guided by a step-by-step improvement procedure. It begins with an initial position and repeatedly improves the current solution by moving to a better one, located in its neighborhood. The procedure continues exploring while the optimal among all feasible solutions (the search space) is not found. The difficulty of the search process is measured by how much time the method consumes to find the optimum. It is principally related to the nature of the search space and how the solving method moves in this space.

This chapter will focus on combinatorial optimization space. It will describe its components, characteristics, and topology. To illustrate that, and how a solution is found by literature search approaches, three problems from three different families are studied. These problems are Traveling Salesman Problem (TSP), JSSP, and QAP. After presenting the search space structure of each COP, common characteristics are highlighted to show how it is relatively easy for resolution methods to navigate from one problem to another, knowing that these COPs are NP-complete.

2.1 Technical Description of Combinatorial Space

For an instance of a given combinatorial optimization problem, a feasible solution S represents a combination, order, permutation, or a set of assignments of its components (variables). All feasible solutions form together the solution or the search space. One or a set of them have the optimal value.

The optimization process is improving continually the initial solution until no improvement is possible, which means a local (possibly the global) optimum is met. The improvement unit is the smallest perturbation performed on the current solution to generate a possibly better new one. The specified perturbation is the step performed to make displacements between solutions and searches in the combinatorial space.

© Springer Nature Singapore Pte Ltd. 2020
A. Ouaarab, *Discrete Cuckoo Search for Combinatorial Optimization*,
Springer Tracts in Nature-Inspired Computing,
https://doi.org/10.1007/978-981-15-3836-0_2

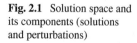

Fig. 2.1 Solution space and its components (solutions and perturbations)

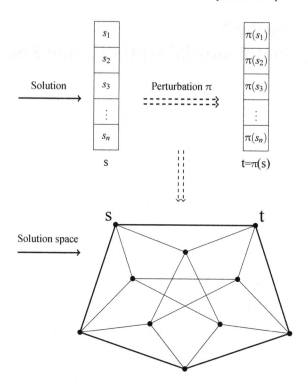

The set S of solutions and π the chosen perturbation are the pair that forms the graph $G = (S, \pi)$ as shown in Fig. 2.1. This graph represents the solution space where S is the set of its vertices and π is the binary relation on S which connects each couple of solutions. The space must be convex, so π must connect each solution, directly or indirectly with all the remainder solutions of the graph. Two solutions are directly connected when we can apply one perturbation (move) to pass from one solution to another or successive moves if they are indirectly connected.

Figure 2.1 illustrates the solution space by the graph composed of vertices that represent a combinatorial solution and arcs that represent perturbations π used to connect solutions. π defines also the space topology, mainly the degree of the graph and the unit distance between two adjacent solutions. When π is changed, the structure of the space is also changed in terms of edges. A unit step is a π that produce the minimum perturbation on the current solution. The move is defined as a partial perturbation on the current solution (position in the space) in order to generate a new one. If the perturbation is minimum, the move is called a small step or a step unit in the neighborhood of the current solution. If the perturbation is important, it is called a big jump out of the neighborhood of the current solution.

To move toward the optimum, an objective function is performed to control movement directions. It is used as a tool to compare the fitness of two solutions, so it indicates where to choose the direction of the next position. It comes after defining the two previously cited components (solution and perturbation). If S represents the structural unit of the space and π indicates how it will be structured, the objective function f defines the movement direction in the space.

Formally, the set of feasible solutions S is defined as follows:

$$S = \{s_1, s_2, \ldots, s_n\}, \tag{2.1}$$

where s_i is the solution variables and n is the problem dimension. A feasible solution is the one which respects the problem constraints. These constraints can be designed as follows:

$$h_j(s) = 0, \ (j = 1, 2, \ldots, J), \tag{2.2}$$

$$g_k(s) \leq 0, \ (k = 1, 2, \ldots, K), \tag{2.3}$$

where h_j and g_k are the constraints related to the problem of interest (Olague 2016).

The fitness of these feasible solutions is measured by the objective function as follows where $s \in S$ is the input and $v \in R$ is the output.

$$f : S \longrightarrow \mathbb{R}^1. \tag{2.4}$$

That is used to define the optimal solution $s^* \in S$ as follows:

$$f(s^*) \leq f(s) \text{ for any solution } s \in S. \tag{2.5}$$

The goal is to determine $s^* \in S$ such that $f(s^*) = \min\{f(s) \text{ where } s \in S\}$. Such a solution s^* is called an optimal solution or a global optimum of the studied problem instance. The following sections provide for each COP a set S of solutions that can be interpreted in the combinatorial space and how the objective function is formulated.

2.2 Studied Combinatorial Optimization Problems

A concrete example of the combinatorial space is illustrated in this section, thanks to three problems from the most studied combinatorial optimization categories. TSP from routing and transportation, QAP from resource allocation, and JSSP from scheduling problem category.

For each problem, a formal definition, solution representation, objective function, and perturbation constraints to keep feasibility of the generated solution are provided. The movement description and its types will be shown in the section related to the solving algorithm and how it searches the space.

2.2.1 Traveling Salesman Problem

The TSP is to find the shortest route or minimum travel cost to visit all n cities exactly once and return to the departure city. A formal definition of TSP according to Davendra (2010) is given as follows:

Let $C = \{c_1, \ldots, c_n\}$ be the set of distinct cities in the space, $E = \{(c_i, c_j) : i, j \in \{1, \ldots, n\}\}$ be the set of arcs between these cities, and $d_{c_i c_j}$ be the cost associated with each arc $(c_i, c_j) \in E$. The TSP consists of finding the minimum length of the closed tour that visits each city once and only once. Cities $c_i \in C$ are presented by their coordinates (c_{ix}, c_{iy}) and $d_{c_i c_j} = \sqrt{(c_{ix} - c_{jx})^2 + (c_{iy} - c_{jy})^2}$ is the Euclidean distance between c_i and c_j. A cycle can be presented a circular permutation $\pi = (\pi(1), \pi(2), \ldots, \pi(n))$ of cities of 1 to n if $\pi(i)$ is interpreted to be the visited city in the step i, $i = 1, \ldots, n$. The cost of a permutation (tour) is defined as

$$f(\pi) = \sum_{i=1}^{n-1} d_{\pi(i)\pi(i+1)} + d_{\pi(n)\pi(1)}. \tag{2.6}$$

If $d_{c_i c_j} = d_{c_j c_i}$, we are talking about the symmetric Euclidean TSP and if $d_{c_i c_j} \neq d_{c_j c_i}$ for at least one arc (c_i, c_j), the TSP is called asymmetric. In what follows, the TSP refers to the symmetric TSP.

Figure 2.2 shows that TSP can be modeled as a weighted graph. It illustrates an example of an optimal solution for the TSP instance "eil51" with 51 (vertices) cities, and the edge weights are the symmetric distances between each couple of cities. The vertices of the graph correspond to the cities, the arcs of the graph correspond to the connections between the cities, the weight of an arc is the distance between its two cities. A tour is a Hamiltonian cycle, and the optimal tour is the shortest Hamiltonian

Fig. 2.2 Optimal tour for TSP instance "eil51"

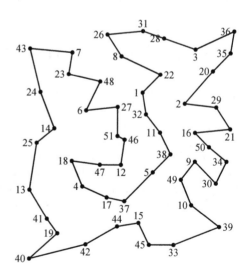

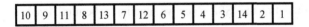

10	9	11	8	13	7	12	6	5	4	3	14	2	1

Fig. 2.3 Optimal solution of "burma14" TSP instance

cycle. Different to the series representation, this representation shows graphically the solution but it is not suitable to be used by solving algorithms. To facilitate the search process, all algorithms applied on TSP use the series representation as shown in Fig. 2.3 that is the optimal solution of TSP "burma14" instance. This representation makes the read/write easier and reduces temporal and spacial complexities.

Perturbation in TSP is defined by a change in the visiting order of two cities at least. It can take many aspects from changing a city order to complex moves (Babin et al. 2007). Generally, to perform a perturbation, TSP does not require specific feasibility constraints.

2.2.2   Job Shop Scheduling Problem

Scheduling problems are a set of combinatorial problems that are commonly defined with regard to the following components:

- A set of operations or jobs to perform,
- A set of resources or machines to use by these jobs, and
- A program to identify, to allocate resources to operations.

Based on these rules, JSSP is formally defined by a set of n jobs $J = \{1, \ldots, n\}$. Each job $j \in J$ must be run on m machines in $M = \{1, \ldots, m\}$ and consists of a sequence of operations (usually m operations) that are executed in a predefined order $O_{j1}, \ldots, O_{jm}$. Following this order, each operation must be performed on a specific $k \leq m$ machine $O = \{O_{ji}, j \in \{1, \ldots, n\}$ and $i \in \{1, \ldots, m\}\}$ for a specific time p_{ji}. Operations are interdependent by two types of constraints. First, an operation O_{ji} can only be started if only the machine to be executed is idle. Second, the priority constraint requires that each O_{ji} operation executes after the completion of its predecessor $O_{j(i-1)}$. Each job must go through each machine once and only once, and each machine can run only one job at a time without interruption.

The objective function of JSSP is formulated while considering the following three notations:

1. O_{ji}: operation i of job j,
2. p_{ji}: runtime of O_{ji}, and
3. C_{ji}: completion time of O_{ji},

where the aim is to find a feasible or admissible order that minimizes the makespan C_{max}, i.e., the time it takes to complete all the jobs ($C_{ji}, i \in \{1, \ldots, m\}$ and $j \in \{1, \ldots, n\}$).

We can then define the JSSP objective function to be minimized, as follows:

$$\text{minimiser } C_{max} = C_{n \times m}. \tag{2.7}$$

The simplest and most used representation for a JSSP instance solution is Gantt chart which associates with each operation a horizontal bar with a length proportional to the time of the operation. Gantt representation gives visual information such as the order of the job operations for each machine. It shows the makespan effective value in the time axis and waiting periods between operations.

As shown in Table 2.1, an example of JSSP instance ("ft06") is presented. It is composed of six jobs and six machines. Column "Jobs" contains the number of the job, and the column "Machine sequence" contains the sequence to be respected when the corresponding job launches its operations. The last column "Processing time" associates with each operation O_{ji} its processing time of the job j on machine i. The optimal solution of this instance is shown in its Gant diagram in Fig. 2.5.

JSSP solution representation is simplified by a sequence of $m \times n$ integers (Fig. 2.4) that is associated with a mod(%n) post-processing function to have values between 1 and n (Fig. 2.6).

JSSP solution is presented in Fig. 2.5 that associates bars with the operations of the jobs. Another representation where bars are associated with the machines is shown in Fig. 2.6 (Esquirol et al. 1999; Herrmann 2006).

2.2.3 Quadratic Assignment Problem

QAP aims to minimize the total cost of construction and operation of facilities knowing that the gain from an economic activity in any site is dependent on other facilities. The QAP solution space is considered as a set of all possible assignments of

Table 2.1 A 6×6 ft06 JSSP instance

Jobs	Machine sequence	Processing times					
0	2 0 1 3 5 4	$p_{00} = 1$	$p_{01} = 3$	$p_{01} = 6$	$p_{03} = 7$	$p_{04} = 3$	$p_{05} = 6$
1	1 2 4 5 0 3	$p_{10} = 8$	$p_{11} = 5$	$p_{12} = 10$	$p_{13} = 10$	$p_{14} = 10$	$p_{15} = 4$
2	2 3 5 0 1 4	$p_{20} = 5$	$p_{21} = 4$	$p_{22} = 8$	$p_{23} = 9$	$p_{24} = 1$	$p_{25} = 7$
3	1 0 2 3 4 5	$p_{30} = 5$	$p_{31} = 5$	$p_{32} = 5$	$p_{33} = 3$	$p_{34} = 8$	$p_{35} = 9$
4	2 1 4 5 0 3	$p_{40} = 9$	$p_{41} = 3$	$p_{42} = 5$	$p_{43} = 4$	$p_{44} = 3$	$p_{45} = 1$
5	1 3 5 0 4 2	$p_{50} = 3$	$p_{51} = 3$	$p_{52} = 9$	$p_{53} = 10$	$p_{54} = 4$	$p_{55} = 1$

2	6	1	8	3	4	5	7	9

Fig. 2.4 An example of a JSSP solution (3×3)

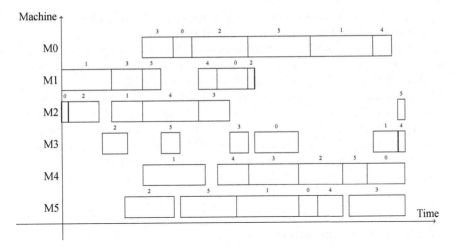

Fig. 2.5 Gantt diagram of the optimal solution of "ft06" JSSP instance

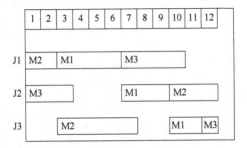

Fig. 2.6 Gantt chart of the solution of Fig. 2.4

Fig. 2.7 Optimum solution of the QAP instance "tai12a"

| 8 | 1 | 6 | 2 | 11 | 10 | 3 | 5 | 9 | 7 | 12 | 4 |

facilities to the given sites. A selected solution S is a permutation ϕ of a given set $Q = \{1, 2, \ldots, N\}$ where N is the dimension of the instance. It also denotes the number of sites and facilities $\phi(i) = k$ means that the facility i is assigned to the site k.

The objective of the problem is to find a permutation $\phi = (\phi(1), \phi(2), \ldots, \phi(N))$ which minimizes

$$\sum_{i=1}^{N} \sum_{j=1}^{N} a_{ij} b_{\phi(i)\phi(j)}, \qquad (2.8)$$

where a is the flow matrix, a_{ij} is the flow between the facilities i and j, and b is the distance matrix. So, the distance from the facility i to the facility j takes the value

$b_{\phi(i)\phi(j)}$. Here, $\phi(i)$ is the location assigned to the facility i. The goal is to minimize the sum of product $flow \times distance$ (Taillard 1991).

Representing graphically QAP instances is not beneficial in terms of analyzing solution fitness and make comparisons; therefore, we will represent it by a series of numbers like the one used for TSP and JSSP instances (Fig. 2.7).

The traveling salesman problem can be considered as a special case of QAP, assuming that the flows connect all facilities on a single ring, and all flows have a constant value greater than zero. Several other problems of combinatorial optimization in the literature are also modeled by the QAP (Ahuja et al. 2000). This shows that designing an approach for solving QAP is a strategic choice that has a positive impact on solving a set of combinatorial optimization problems.

2.3 Common Characteristics

As known, combinatorial optimization problems are grouped in three main classes, considering their complexity that depends on the corresponding decision problem, i.e., a problem that answers the question: is there a better solution than the current one?

The first class is "P class" (Polynomial time). It brings together all the decision problems that can be solved in polynomial time proportionally to the size of the entry. We can consider the problems of this class as so-called "doable" problems.

The second class is the class of decision problems treated in polynomial time, by a non-deterministic algorithm. It is called "class NP". Following this definition, we can say that the class P is included in the class NP. We also cite that the equality of these two classes is a problem not yet solved. The problem $P = NP$ is considered one of the most studied problems in theoretical computer science (Cook 2000).

The last class is the "NP-complete class", which contains the problems of class NP such that any problem of NP is reducible to them in polynomial time. We say that a combinatorial optimization problem is NP-hard if its corresponding decision problem is NP-complete (Arora and Barak 2009). A very large number of real-world optimization problems are NP-hard, and therefore any progress in dealing with NP-complete problems will have a very significant impact on many applications.

NP-complete is a class of problems that are characterized by the fact that if for one of its problems there exists a way to solve it, the whole class will be solved. Therefore, studying how NP-complete problems are connected can serve resolution algorithms. To serve this purpose, this section will show how these problems are connected and what are their common properties. It will focus on two of their sides: similar representation of structure and constraints of adapting the perturbation (Fig. 2.8).

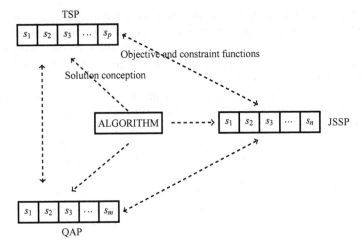

Fig. 2.8 Changing problems solved by one algorithm

2.3.1   Representation

As shown in Ouaarab and Yang (2016), TSP, JSSP, and QAP are solved by cuckoo search using similar representations. It explains how one approach can propose a unique model to solve a variety of problems. For these three problems, a series of numbers is used to facilitate operations and search procedures in the solution space. The algorithm performed one strategy to search in the three solution spaces. It adapts just parts related to moves plugged by the problem and feasibility constraints.

2.3.2   Constraints

Passing from one problem to another is changing the interpretation of its representation to the new one. Interpreting the sequence that represents the problem is defining the new move applied on the sequence with respect to the objective function. So, the algorithm adapts, just, its moves to generate a feasible solution regarding the new problem constraints, and fitness is returned by the problem objective function.

2.4   Conclusion

Because of their NP-hard complexity, analyzing the conception of combinatorial optimization problems becomes very important in order to gain more computational time. This is why methods, developed to solve these problem categories, instead of just developing resolution strategies, propose in addition their adapted conception for the treated problem in the hope that this conception reduces the complexity. The conception of a combinatorial problem considers the solution representation, perturbation or move in solution space, and the objective function to evaluate the fitness of a solution and the gain of a move.

To validate a newly developed method using NP-hard problems, the choice of the well-adapted problem conception is crucial. This way, the method enhances the chances to prove its performance. This is why a proposed design for a given problem decides what methods will solve it. And then, it can be adapted (with a reduced changes) to solve an another problem in the same class of complexity.

The purpose of this chapter is to show that the conception of NP-hard combinatorial optimization problems is very important as much as their resolution. The aim of optimization methods is not finding a solution, but finding the best in the minimum duration, and time can be reduced if the problem is well designed.

References

Ahuja RK, Orlin JB, Tiwari A (2000) A greedy genetic algorithm for the quadratic assignment problem. Comput Oper Res 27(10):917–934

Arora S, Barak B (2009) Computational complexity: a modern approach. Cambridge University Press

Babin G, Deneault S, Laporte G (2007) Improvements to the Or-Opt heuristic for the symmetric travelling salesman problem. J Oper Res Soc 58(3):402–407

Cook S (2000) The p versus np problem. In: Clay Mathematical Institute; the millennium prize problem. Citeseer

Davendra D (2010) Traveling salesman problem, theory and applications. InTech

Esquirol P, Lopez P et al (1999) L'ordonnancement. Economica

Herrmann JW (2006) Handbook of production scheduling, vol 89. Springer Science & Business Media

Olague G (2016) Evolutionary computer vision: the first footprints. Springer

Ouaarab A, Yang X-S (2016) Cuckoo search: from cuckoo reproduction strategy to combinatorial optimization. In: Nature-inspired computation in engineering. Springer, pp 91–110

Taillard E (1991) Robust taboo search for the quadratic assignment problem. Parallel Comput 17(4):443–455

Chapter 3
Solving Combinatorial Optimization Problems

3.1 Combinatorial Optimization Problem (COP) Resolution Approaches

As known, most of the combinatorial optimization problems are NP-hard in terms of complexity and they are solved as part of one of the three predefined classifications: solution construction, solution improvement (or trajectory algorithms), and population-based metaheuristics. It is also known that it is practically very difficult to have both an optimal solution quality and a reduced computation time. Indeed, most conventional algorithms make the choice between a high quality of the solution and an exponential computation time, or a solution of modest quality and a polynomial time. The third choice offers a good (not necessarily optimal) solution in a reasonable computation time.

This chapter is limited to the population-based metaheuristics. It presents two aspects of how population-based metaheuristics use their individuals while searching for the optimum. Each individual of the population behaves in relation to the rest of the population. Therefore, we have grouped these methods into two classes: vertical improvement that manages the information shared over individual generations and controls the information flows between individuals of the same population in horizontal improvement.

Considering the aforementioned solution space structure, for each class of algorithms, we will highlight the following main aspects:

- Intensification/diversification (I/D): How algorithms intensify their search in the promising region near to a potential optimum and diversify the search to escape from the local optimum.
- Population structure: How individuals of the population are performed; is there a variety of them or they are identical; what kind of relation between them?

While passing pertinent information, vertical and horizontal improvement metaheuristics try to control their convergence rate by performing strategies to move in

© Springer Nature Singapore Pte Ltd. 2020
A. Ouaarab, *Discrete Cuckoo Search for Combinatorial Optimization*,
Springer Tracts in Nature-Inspired Computing,
https://doi.org/10.1007/978-981-15-3836-0_3

the solution space. So they will be studied with focus on the I/D, population structure, and stochastic strategy.

3.1.1 Vertical Improvement

To well describe the vertical improvement by a concrete example, we can propose Genetic Algorithms (GA) that are a part of the evolutionary algorithm family which are inspired in principle by the concept of natural selection developed by Darwin (1998). They are designed to simulate a population of individuals, on which it applied two principal operators (crossover and mutation) and are then subject to a selection for making the next generation. The vertical improvement over generations is guided by the objective function of the treated problem.

3.1.1.1 Genetic Algorithm

In 1975, John Holland published his book "Adaptation in Natural and Artificial Systems" (Holland 1975) to highlight and explain rigorously the processes of adaptation of natural systems and to design artificial systems that use these mechanisms. Its genetic algorithm consists in changing a population of chromosomes (represent the solutions of the studied problem) to a new population using a natural selection and then applying the operators inspired by genetics in this case crossbreeding and mutation. It maintains a population of individuals for each generation. Some individuals (parents) of the population are selected and transformed with genetic operators to generate a new population (children). Each individual is evaluated by a fitness function. The process will stop once a solution is found or when a maximum number of generations has been reached.

Genetic algorithms represent the population individuals by chromosomes that requires to be encoded, for the objective function, with binary arrays or strings. Chromosomes are manipulated by genetic operators, and selected according to their aptitude. The key steps are done by the following procedure (Yang 2010):

- Coding of the objective (optimization) function.
- Define the selection criterion.
- Initial population of individuals.
- Evaluate the fitness of all individuals in the population.
- Create a new generation by crossing and mutation.
- Evolution of the population until the verification of the stopping criteria.
- Decoding the result to get the solution of the problem.

Genetic algorithms make improvement by starting with an initial population and then a portion of the best individuals are selected to pass their promising qualities to the next generation. This way, GA improves vertically their individuals over gen-

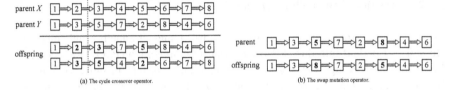

Fig. 3.1 GA principal moves

erations, and no information is shared between individuals of the same generation. The only improvement flow is vertical.

As discussed, improvements are performed by moves from one position to a new better one. In GA, the move is from parent generation to his child generation. It is realized, thanks to crossover and mutation operators. An example of applying Crossover (CX) and mutation (swap) operators is shown, respectively, in Fig. 3.1a and b. Crossover generates an offspring (descendant) that possesses the common best qualities of its parents (ascendant) while mutation is a generation of an offspring that has partially the same qualities as his parent with a portion of new qualities.

Crossover and mutation operators are used to balance search in the solution space while running intensification and diversification mechanisms. Intensification consists in pushing research into a subspace of individuals that looks promising, while diversification makes it possible to move on and discover new regions of solution space. We can consider that the mutation operator makes rather exploration by diversifying the search from the current parent to a child with new characteristics. Crossover is an exploitation of a region (offspring) between two promising regions (selected parents).

In theory, crossover is a pertinent method to keep search between the two parents regions. However, it is not simple to realize this concept when the treated problem is combinatorial. As shown in Fig. 3.1a, the generated offspring is not completely composed of its parents' parts. It needs a post-rectification to become feasible. Mutation is performed by the predefined moves of the problem. If the move is a small step, the mutation is considered as an intensification and diversification with big steps.

3.1.2 Horizontal Improvement

In horizontal improvement, information is shared between individuals of one generation. They benefit from each other experiences. Three examples are given to illustrate how metaheuristics propose different strategies to horizontally improve their population.

3.1.2.1 Particle Swarm Optimization

In 1995, the American social psychologist James Kennedy and engineer Russell C. Eberhart introduced Particle Swarm Optimization (PSO) (Kennedy and Eberhart 1995). Roughly, PSO is an optimization algorithm inspired by fish and bird swarm intelligence and even human behavior (Yang 2010). The principle is to start with a population of random solutions and search for an optimum by updating the generations, and move in the solution space by considering the current optimum. Each particle keeps track of its coordinates, in the search space, which feeds particle Best (pBest) the best solution achieved so far. The other best solution local Best (lBest) followed by the particles is the best solution obtained so far by all the neighboring particles. When a particle takes the whole population as its topological neighbors, the best value is a better overall solution global Best (gBest).

Particle swarm optimization is a stochastic optimization technique using a population of particles. It is based on a global communication between the particles (individuals). Each particle moves with a variable speed toward its own best position pBest in the past and the best global position gBest according to Eq. 3.1:

$$X_{n+1}^{(i)} = X_n^{(i)} + V_{n+1}^{(i)}, \qquad (3.1)$$
$$V_{n+1}^{(i)} = wV_n^{(i)} + c_1 r_1 (pBest_i - X_n^{(i)}) + c_2 r_2 (gBest - X_n^{(i)}), \qquad (3.2)$$

where, in the iteration $(n + 1)$, X^i $(i = 1, 2, \ldots, m)$ is the ith among m particles, which move with the velocity V^i. P_i and P_g are, respectively, $pBest$ and $gBest$. w, c_1, and c_2 represent the parameters that are selected, in the interval $[0, 1]$, to control the behavior of PSO. r_1 and r_2 are randomly generated in $[0, 1]$. The velocity decreases when the particle is brought closer to $pBest$ and $gBest$.

To understand how PSO makes the balance between intensification and diversification, we will see the effect of increasing and decreasing each parameter in Eq. 3.1. Intensification is divided between two regions: the best found position of each particle and the global best. It is controlled by c_1 and c_2. If c_1 is big enough, the particle prefers to move toward the region of its historical best position and the search is intensified around the global best position if c_2 is bigger than c_1. r_1 and r_2 add the exploration aspect to the search process. If their values is 0, there will be no exploration in the path of the particle toward $pBest$ and $gBest$; on the other hand, the random exploration is important when r_1 and r_2 are big.

To introduce the concept of velocity into a combinatorial space, several attempts have succeeded in proposing an adaptation of PSO to a set of combinatorial optimization problems. But, during the transition from continuous space to combinatorial space, these adaptations are incapable of keeping the properties between the two spaces. This constraint is always encountered in all adaptations of distance, velocity, or displacement in continuous space with respect to combinatorial one. In order to circumvent this constraint, several alternatives are studied, like the approach of Shi et al. (2007), based on "Swap operator" developed by Wang et al. (2003). PSO proposes for each treated problem at least one move to guide a combinatorial velocity.

3.1.2.2 Firefly Algorithm

Every night, the sky of the tropical temperate regions is animated by a spectacle of the flashing lights of the "fireflies". They are about two thousand species and most produce short, rhythmic blinks. Each particular species often has its own pattern of blinking. The functions of these signaling systems still remain a debate. However, two basic functions of such flashes are to attract mating partners and potential prey. On the other hand, they can also serve as warning mechanisms. Females respond to a single pattern of blinking of a male of the same species, while in some species such as "Photuris", female fireflies can mimic the pattern of mating flicker of other species in order to attract and eat the male fireflies that can be considered a suitable companion potential (Yang 2009a).

Firefly Algorithm (FA) developed by Xin-She Yang in 2008 is inspired by the flash pattern and firefly characteristics. To simplify the description of the algorithm, he uses the following three idealized rules (Yang 2009b):

- All fireflies are unisex so that fireflies will be attracted to other fireflies, regardless of their gender;
- The attractiveness is proportional to the brightness, so for two flashes of fireflies, the one that is less luminous will move toward the brightest. Attractiveness is proportional to brightness and both decrease as distance increases. If there is no other light source, the firefly will move randomly;
- The brightness of a firefly is affected or determined by the topology of the objective function. For a maximization problem, the brightness can be simply proportional to the value of the objective functions.

In FA, there are two important issues: the variation of light intensity and the formulation of attractiveness. For simplicity, we can always assume that the attractiveness of a firefly is determined by its brightness, which is in turn associated with the coded objective function.

Intensification and diversification appear in the behavior of fireflies, respectively, in attractiveness and these random displacements. The intensification takes place, in a first level, by the intensity of the light which limits the movement of the fireflies toward the most luminous source, and a second level by the distance of the light source and the absorption of light which decreases the field of visibility. Diversification is achieved by a random behavior of fireflies, which partially ignores the flashes produced by others.

The balance between intensification and diversification is done using the following formula which describes the motion of a "i" firefly at the "x_i" position toward the brightest firefly "j" in the "x_j" position:

$$x_i = x_i + \beta_0 e^{\gamma r_{ij}^2}(x_j - x_i) + \alpha \varepsilon_i, \tag{3.3}$$

where r_{ij} is the Cartesian distance from x_i to x_j, γ is the absorption coefficient of light, β_0 is the attractiveness in $r = 0$, ε_i is a random number vector from a distribution, and α is the randomization parameter.

3.1.2.3 Bee Colony Optimization

Algorithms inspired by the Bee Colony Optimization (BCO) have begun to emerge in the last 10 years as a promising and powerful tool. They were developed in a few years independently by several groups of researchers. In 2004, Nacrani and Tovey formulated honey bees to give computers between different clients on the web hosting servers. Around the year 2006, Basturk and Karaboga developed an Artificial Bee Colony (ABC) algorithm for the optimization of digital functions (Basturk and Karaboga 2006).

The aim of the bee colony is to optimize the overall efficiency of nectar collection. Foragers are assigned to different food sources to maximize total nectar intake. The distribution of bees is a function of several factors such as the richness of the nectar and the proximity of the hive. The most important point in the dance of bees is their strength to attract the attention of others. In the colony, bees are divided into three groups:

- bees employed (foragers),
- bees spectators (observers), and
- scouting (explorers).

For each food source, there is only one bee employed. The number of bees employed is equal to the number of sources of food. The bee employed from an abandoned food source is forced to become a "scout" in search of new sources of food at random. The employed bees share information with the bee spectators so that the spectators have the opportunity to choose a food source. Unlike the bee algorithm that has two groups of bees (foraging and observing bees), bees in ABC are more specialized (Yang 2010).

After locating a food source, the bee returns to the hive and initiates the communication process, in the form of dances, with the rest of the hive. Bees that observe have the choice to follow the clues provided by this dance or to ignore them. We have, therefore, in principle two factors that affect the foraging: the weight of the dance and the probability of ignoring the dance and go to a random search.

These two factors (the strength of the dance and the probability of a random exploration of space) are presented in a simple way by Quijano and Passino in the

"Honey bee" (Quijano and Passino 2010) version by Eq. 3.4 of probability p_i of an observing bee.

$$p_i = \frac{w_i^j}{\sum_{i=1}^{n_f} w_i^j},$$ (3.4)

where w_i^j is the weight of the dance of the bee i in the iteration $t = j$, and n_f is the number of bees in foraging. The number of observing bees is $N - n_f$ where N is the total number of bees. The probability of exploration is defined as a Gaussian expression, as follows:

$$p_e = 1 - p_i = e^{-\frac{w_i^2}{2\sigma^2}},$$ (3.5)

such as σ (fixed) is the volatility of the colony.

Intensification is guided by p_i. If it is important, the observer will follow the forager that has big w_i relative to the others. If not, the dances are not interesting to the observers, and the number of explorers will be increased. Over iterations, less good positions are found which implies less interested observer bees to the dances. This situation enhances the number of explorers, so the exploration rate and the intensification appear when more promising regions are found.

3.2 Discretization

The great part of metaheuristics such as those presented in this chapter try to solve combinatorial optimization problems after an adaptation to the problem solution space. Adaptation mechanisms allow metaheuristics to search and move in the combinatorial space, but their robustness still not as good as searching in the continuous space.

In this section, we will show how algorithms (metaheuristics) behave with COPs in three different ways. Originally, discrete algorithms designed initially for the combinatorial space, discreted algorithms that are adapted to search in the combinatorial space and continuous algorithms that keep their continuous search strategies but uses an intermediate that projects their displacement on the combinatorial space.

3.2.1 Discrete

Discretely designed metaheuristics are considered as algorithms that search without adaptation in the combinatorial space. Two examples are given in this subsection: Ant Colony Optimization (ACO) and genetic algorithm.

According to Dorigo and Birattari (2010), the basic version of the ACO is designed to solve the traveling salesman problem (Dorigo et al. 1997). When an ant begins with a random move, from a city r to s, passing through the arc that connects these two cities, it deposits a quantity of pheromone. The other ants that follow this ant will have the choice to consider the deposited pheromone and pass on this arc or to take an arc randomly according to the following formula:

$$
p_k(r, s) = \begin{cases} \dfrac{[\tau(r, s)] \cdot [\eta(r, s)]^\beta}{\sum_{u \in M_k} [\tau(r, u)] \cdot [\eta(r, u)]^\beta} & \text{if } s \notin M_k \\ 0 & \text{otherwise} \end{cases}, \tag{3.6}
$$

where $p_k(r, s)$ is the probability of choosing (by the ant k) the passage from the city r to the city s, M_k is the set of cities visited and stored by the ant k, $\tau(r, s)$ is the amount of pheromone deposited on the arc (r, s), $\eta(r, s)$ is the heuristic function, which can be the inverse of the distance between cities r and s, and β is a weight parameter relative to the importance of the arc, used by ACO to promote the arc of the best solutions, with centralized control. On the other hand, bad decisions in the past are forgotten during iterations with the evaporation of pheromone (Colorni et al. 1991).

ACO is different in terms of generating solution compared to all cited metaheuristics of this book. Ants move to construct a solution rather than search in a space of solutions. It moves naturally, but over the components of a routing problem solution.

The second example is GA. It represents individuals by chromosomes composed of a binary series. With this representation, GA can be considered as an initially discrete algorithm but not for all combinatorial problems. It needs adaptation for those that have precedence or order constraints, especially after applying the crossover operator. It generates redundant numbers that must be fixed with a feasibility post-process.

3.2.2 Discreted

Particle swarm optimization, firefly algorithm, and bee colony optimization are designed to move in the continuous space. Their equations contain variables with, naturally, real values. In the case of combinatorial problem, these variables will be transformed by discretization mechanisms (Ouaarab and Yang 2016). Discretization process considers mainly solution feasibility and I/D.

3.2.2.1 Feasibility

To keep generating feasible combinatorial solutions, metaheuristics have to consider in the case of TSP, JSSP, and QAP the following constraints:

- TSP: Each city must appear once;
- JSSP: Tasks respect the precedence constraint and machine order of execution;
- QAP: Bijective assignment between facilities and sites.

These constraints must be respected in the generation of initial population and while moving in the space after perturbing solutions.

3.2.2.2 I/D

All metaheuristics try, while adapting the search, to bring their I/D strategy robustness into the combinatorial space. They have to focus in realizing intensification by small steps around the optima and diversification by big jumps far from the current solution. Therefore, at least one small and big step has to be proposed to allow local and global searches.

3.2.3 Projected

In the projected case, metaheuristics do not need any adaptation. They search in continuous space with their own strategies. The adaptation is in the form of a projection (Ouaarab et al. 2015). The generated continuous solution is projected in the combinatorial space. This way, metaheuristics keep their performance and do not have to worry about combinatorial concerns. This notion will be detailed in the last chapter of this book with experimental examples.

3.3 Conclusion

A great part of metaheuristics, which are generally population-based algorithms, are inspired by the collective behavior of groups, colonies, or swarms of several species in nature. These collective behaviors are adopted by swarms in order to find a partner to produce their next generation, a food source, or avoid predators over a large space, using relatively simple tools for sharing experiences and useful information. Therefore, two strategies can be adopted to find the optimized solution: the first (horizontal) is improving the individual fitness by sharing information, over the swarm, about the search space, and the second (vertical) is producing a new generation of better individuals that are capable to search more efficiently. This is what this chapter shows with examples in the first section.

In the second section, it divides the adaptation methods into three strategies. Discrete metaheuristics have limited choices of moves and designed for a specified category of problems, but they are implemented easily. Discreted metaheuristics have unlimited choices of moves, and the adaptation mechanism is applied on a variety of combinatorial problems. The third tool of adaptation let the metaheuristic search freely in the continuous space, and then the found solutions are projected in the combinatorial space.

References

Basturk B, Karaboga D (2006) An artificial bee colony (abc) algorithm for numeric function optimization. In: IEEE swarm intelligence symposium, pp 12–14

Colorni A, Dorigo M, Maniezzo V (1991) Distributed optimization by ant colonies. In: Proceedings of the first European conference on artificial life, vol 142. Paris, France, pp 134–142

Darwin C (1998) The origin of the species

Dorigo M, Birattari M (2010) Ant colony optimization. In: Encyclopedia of machine learning. Springer, pp 36–39

Dorigo M, Gambardella LM et al (1997) Ant colonies for the travelling salesman problem. BioSyst 43(2):73–82

Holland J (1975) Adaptation in natural and artificial systems

Kennedy J, Eberhart R (1995) Particle swarm optimization. In: IEEE international conference on proceedings neural networks, 1995, vol 4. IEEE, pp 1942–1948

Ouaarab A, Ahiod B, Yang X-S (2015) Random-key cuckoo search for the travelling salesman problem. Soft Comput 19(4):1099–1106

Ouaarab A, Yang X-S (2016) Cuckoo search: from cuckoo reproduction strategy to combinatorial optimization. In: Nature-inspired computation in engineering. Springer, pp 91–110

Quijano N, Passino KM (2010) Honey bee social foraging algorithms for resource allocation: theory and application. Eng Appl Artif Intell 23(6):845–861

Shi X, Liang Y, Lee H, Lu C, Wang Q (2007) Particle swarm optimization-based algorithms for tsp and generalized tsp. Inf Process Lett 103(5):169–176

Wang K-P, Huang L, Zhou C-G, Pang W (2003) Particle swarm optimization for traveling salesman problem. In: 2003 international conference on machine learning and cybernetics, vol 3. IEEE, pp 1583–1585

Yang X-S (2009a) Firefly algorithm, lvy flights and global optimization, pp 209–218

Yang X-S (2009b) Firefly algorithms for multimodal optimization. Stochastic algorithms: foundations and applications, pp 169–178

Yang X-S (2010) Engineering optimization: an introduction with metaheuristic applications. Wiley, USA

Chapter 4
Cuckoo Search: From Continuous to Combinatorial

When solving combinatorial optimization problems with nature-inspired metaheuristics (which are mostly initiated in continuous spaces), the first constraint to manage is how to move in the combinatorial space of solutions without affecting the performance of these metaheuristics. The definition of the neighborhood of a solution in space represents a second constraint.

The ability to balance research between the exploitation of promising areas and the exploration of the search space on a global scale is a performance parameter linked to all metaheuristics. In the search process for the optimal solution, the biggest inconvenience is to be blocked by a local optimum. The only way to escape from the local optimum region is by performing a big exploring jump to another distant region. But jumping between regions is not sufficient to find the best solution. Metaheuristics have, for this reason, to risk intensifying the search. Therefore, the most robust metaheuristic is that which balances perfectly between intensification and diversification (Intensification/diversification (I/D)). In this chapter, the main notions that will be discussed are related to I/D. Considering this purpose, Cuckoo Search (CS) is described.

Since its appearance in 2009, CS has been cited more than 4400 times. The most important is not the number of citation, but the diversity of the application of this algorithm. This confirms our theory about its quality of being generic and easily adaptable to different types of problems in a wide variety of application domains. The use of optimization methods to solve design problems in the field of engineering (Design Optimization) remains very important, and it is attacked more and more recently by the metaheuristics. Optimization of designs is an integrated part of product design engineering and industry. Most of the design problems are considered complex, and there are several objectives to optimize, and sometimes the optimal solution is not yet found. In this type of problems, CS has been applied, considering the structures "design of springs" and "welded beam". The best solution found by CS far surpasses that found by an efficient "particle swarm optimization" (Yang and Deb 2010). A new version of CS for multi-objective optimization is formulated. This new

© Springer Nature Singapore Pte Ltd. 2020
A. Ouaarab, *Discrete Cuckoo Search for Combinatorial Optimization*,
Springer Tracts in Nature-Inspired Computing,
https://doi.org/10.1007/978-981-15-3836-0_4

version aims to solve the multi-objective problems that are typically different from single-objective optimization problems. The difference between these two types of problems lies mainly in the effort of calculations and the number of objective functions which increases considerably (Yang and Deb 2013). CS is also a solution in the field of optimization and applications for companies (Yang et al. 2012). It is also proposed to solve the problems of optimization of truss structures (Gandomi et al. 2012) and several other problems of structures optimization (Gandomi et al. 2013). Another interesting area is software engineering, where CS can offer solutions. "Test effort estimation" is a procedure for predicting the effort to test software and what is the time needed to test a given system (Srivastava et al. 2012). For more examples, we can find a set of several applications, variants, and improvements of CS in Yang and Deb (2014) and Fister et al. (2014).

CS has a particular strategy to make controlling the balance between local intensive search strategy and a more efficient exploration of the entire search space. Also, the reduced number of parameters makes CS a less complex algorithm and therefore potentially more generic. This is the result of using a small number of control parameters. Indeed, CS uses two parameters, the population size n and the portion of abandoned nests p_a. In principle, n is fixed and p_a essentially controls the elitism and balance between randomization and local search. The reduced number of parameters makes the algorithm less complex and therefore potentially more generic. All this is explained by the ease of tuning both CS parameters while keeping its robustness to solve a wide range of optimization problems even the qualified problems are NP-hard (Yang and Deb 2009).

In this chapter, we will present a description of the algorithm we took as a model for solving combinatorial optimization problems such as Traveling Salesman Problem (TSP), Job Shop Scheduling Problem (JSSP), and Quadratic Assignment Problem (QAP). We will begin with a description of CS and its formulation, from inspiration source to its improved version which is an improved search process and a new population structure. The third section of the chapter talks about the design of discrete CS (Discrete Cuckoo Search (DCS)) starting from improved CS. It dissects CS into four main components: nest, egg, objective function, and search space. During the adaptation of DCS, these main notions will appear in the text: egg, nest, objective function, search space, and Lévy flights.

4.1 Cuckoo Search Description

Cuckoo bird is discreet elongated, medium size, and its song marks the beginning of the beautiful season and its fascinating reproduction strategy. Old world and some new world cuckoos practice inter-species brood parasitism by laying one or more eggs in a previously observed host nest. The study of Yang et al. (2017) shows that cuckoo parasitism is affected mainly by the host activities instead of host nests. Therefore, cuckoo females look for host birds that guided by their natural instinct of hatching, brooding, and bringing food to the little cuckoos, and so increase their probability of passing to the next generation and avoid abandoning their egg by the host bird.

Brood parasitism behavior of cuckoos is combined in CS with Lévy flights to effectively search for a new nest. The Lévy flights, named by the French mathematician Paul Lévy, represents a model of random walks characterized by their step lengths that follow a distribution of power law that is written from Eq. 4.1 (Bak 1997):

$$N(s) = s^{-t}. \tag{4.1}$$

In nature, animals search for food randomly or almost randomly. In general, the search for food is actually a random walk because the next move is based on the current location/state and the probability of transition to the next location. The direction they choose depends implicitly on a probability that can be modeled mathematically. Various studies have shown that the flight behavior of many birds and insects has the typical characteristics of Lévy flights (Yang and Deb 2010). On the other hand, Reynolds and Frye (2007) have shown that fruit flies or "Drosophila melanogaster" explore their landscape using a series of straight trajectories punctuated by a sharp turn to 90°, which leads to an intermittent search pattern without scale of a Lévy flights' style. This model is commonly represented by small random steps followed in the long run by large jumps (Brown et al. 2007; Shlesinger et al. 1995; Yang and Deb 2009). These probability distributions are known as Lévy distributions or stable distributions. The lengths, l, of the steps or of the jumps of the March are distributed according to the power law, $P(l) = l^{-\mu}$ with $1 < \mu \leq 3$ (Viswanathan et al. 1999).

CS is a metaheuristic, developed recently by Xin-She Yang and Suash Deb in 2009, originally designed to solve multimodal functions. It takes as a base the following ideas:

- Each cuckoo lays one egg at a time and chooses a nest randomly;
- A good nest of good quality can move to a new generation;
- The number of host nests is fixed, and a cuckoo's egg can be discovered by the bird host according to a probability $p_\alpha \in [0, 1]$.

A cuckoo i generates a new solution $x_i^{(t+1)}$ via the Lévy flights, according to Eq. (4.2)

$$x_i^{(t+1)} = x_i^{(t)} + \alpha \oplus Levy(s, \lambda), \tag{4.2}$$

where α is the length of the step following the Lévy flights distribution shown in Eq. (4.3):

$$Levy(s, \lambda) \sim s^{-\lambda}, \quad (1 < \lambda \leq 3), \tag{4.3}$$

which has infinite variance and mean (Yang and Deb 2009). Here, s is the size of the step taken from the Lévy distribution.

The cuckoo search algorithm is based on the following ideal rules:

- Each cuckoo egg in a nest represents a solution.
- Each cuckoo bird will lay a single egg at a time and will choose a nest "*randomly*". Therefore, every individual in the cuckoos' population has the right to randomly generate a single new solution.
- The best nests of better quality of eggs will lead us to the new generations. Here, we implicitly introduced the notion of intensification or research around the best solutions.
- Some new solutions must be generated by Lévy flights around the best solution obtained so far. This will speed up the local search.
- The number of host nests is fixed, and the egg laid by the bird is discovered by the host with a probability $p_a \in [0, 1]$. In this case, the host bird chooses to get rid of the egg, or abandon the nest and rebuild another nest somewhere. For simplification, this last assumption will be approximated by the p_a fraction of the n nests that are replaced by new ones (new random solutions).
- A significant fraction of the new solutions must be generated by randomization to remote areas and whose locations must be far enough from the current best solution, which will ensure that the system will not be trapped in a local optimum.
- Each nest can contain several meaningful eggs in a set of solutions.

These rules are formulated in Algorithm 1. It is composed of three main procedures:

- Firstly, the initial population of n nests is generated. In general, it is generated randomly or by using a pre-processing method to accelerate the convergence;
- Then, among the initial population, a randomly selected individual is compared to a cuckoo in the solution space. The better of them takes a place in the population;
- The third procedure launches an exploration of the search space for new solutions which will replace a fraction of p_a abandoned worst nests.

Algorithm 1: Cuckoo Search

1: Objective function $f(x), x = (x_1, \ldots, x_d)^T$
2: Generate the initial population of n nests $x_i (= 1, 2, \ldots, n)$
3: **while** (t <MaxCeneration) or (stop criteria) **do**
4:     Get a cuckoo randomly via Lévy flights
5:     Evaluate its quality/fitness F_i
6:     Choose a nest among n (j) randomly
7:     **if** ($F_i < F_j$)(minimization) **then**
8:         Replace j by i
9:     **end if**
10:     a fraction (p_a) of worst nests is abandoned and new ones are built
11:     Hold the best solutions (or nests with high-quality solution)
12:     Classify the solutions and find the current best
13: **end while**
14: Post-process of results and visualization

The search for a new random solution $X^{(t+1)}$ via Lévy flights is accomplished by the formula below:

$$X^{(t+1)} = X^{(t)} + \alpha \oplus Levy(\gamma). \tag{4.4}$$

This equation is stochastic (a random walk). In general, a random walk is a Markov chain whose next step depends only on the current step, which is the first term of the equation, followed by the transition probability which is the second term. The product $\oplus$ represents the matrix product. α is the maximum length of the step that should be related to the search space scale of the problem. In that case, $\alpha = 1$. Random walk through Lévy flights is more effective in exploring the search space than other random walks, as its step size is much larger in the long run.

CS is a population-based algorithm, which gives it some similarity to genetic algorithms or particle swarm optimization, except that it uses a kind of elitism and/or selection similar to that used in "harmony search" (Geem et al. 2001). Considering our classification of horizontal and vertical improvements, CS brings the two together. It improves its population individuals (cuckoos) vertically from one to the next generation by selecting which cuckoo will pass to the next generation. Replacing cuckoos is a horizontal improvement which is performed in two operations. The first is replacing a random individual by a better cuckoo from the solution space, and the second is the replacement of the portion of worst cuckoos.

Cuckoo search algorithm has two capabilities: a local search and a global search, controlled by its switch/probability of discovery parameter. Overall search is favored by p_a which takes the value $1/4$ (0.25), while local search is boosted in $3/4$ of the population. This allows the CS to proceed efficiently while balancing between exploration and exploitation. Another advantageous factor is the choice of Lévy flights, instead of random walks. We cannot, therefore, neglect the reduced number of parameters that allows CS to be more generic. In addition to its benefits, we have found that CS has more room for more improvements, either in terms of its source of inspiration or in the algorithm itself. There is also randomization which is more effective if the step size is larger in the long run. Finally, the number of parameters to be adapted is smaller than Genetic Algorithms (GA) and Particle Swarm Optimization (PSO), and therefore more generic to adapt to a large class of optimization problems.

The strength of CS is how it uses and explores the solution space with cuckoos. The cuckoo shows some intelligence in order to detect the best solutions. So, it represents a direct control tool to intensify or diversify searches.

4.2 Improved CS

The robustness of CS is based on how to explore and exploit the space of solutions by a cuckoo. This cuckoo can have some "intelligence" to find much better solutions. We considered in our improvement (Ouaarab et al. 2014) the cuckoo as the first level of control of intensification and diversification, and since this cuckoo is an individual of a population, then this population is qualified to be the second level of control. The idea of improvement is to restructure the population by integrating a new, relatively smarter category of cuckoos with more efficiency into their research, compared to other cuckoos.

Studies have shown that the cuckoo is able to initiate surveillance around potential host nests (Payne and Sorenson 2005). This behavior can serve as an inspiration to design a new cuckoo category that has the ability to change host nests during incubation. The purpose of this behavior is to avoid abandoning cuckoo eggs. These cuckoos adopt mechanisms before and after brooding. They observe the chosen host nest to be sure that the choice of this nest is the right decision or not (in this case, they start looking for a new choice much better than the current one). We speak, therefore, of an ability to search locally for a much better solution around the current solution.

Inspired by this observed behavior, the mechanism adopted by this new fraction of cuckoos can be divided into two main stages: (1) A cuckoo initially moves by Lévy flights to a new solution (which represents a new region); (2) From the current solution, a cuckoo in the same region is looking for a new best solution (in this stage, it is possible to carry out a local search). According to these two steps, the population of the improved CS algorithm (Algorithm 2) can be structured according to three main categories of cuckoos:

1. A cuckoo looking for (from the best position) regions that may contain new solutions that are much better than the solution of an individual randomly selected in the population;
2. A p_a fraction of cuckoos looking for new solutions far from the best solution;
3. A p_c fraction of cuckoos looking for solutions from the current position and trying to improve them. They move from one region to another via Lévy flights to locate the best solution in each region without being trapped by the local optimum.

We note that the cuckoo population, during its search process, is guided by the best solution, the solutions found locally, and the solutions found far from the best solution. This improves the intensive search around different best solutions, and at the same time, a randomization is properly carried out to explore new regions using Lévy flights. Thus, an extension of the standard version of CS, as shown in bold in Algorithm 2, is the addition of a method that manipulates the fraction p_c of intelligent cuckoos. This allows CS to operate more efficiently with fewer iterations and show more resistance to potential stagnations in local optima.

The new process added to standard CS can be illustrated by a procedure considering that the value of the fraction p_c is set to 0.6, such that this fraction is a set of good solutions of the population except the best solution. Starting from each solution, a cuckoo searches randomly through Lévy flights around the current solution and then tries to find the best solution in this region using local search.

The aim of this improvement is to strengthen intensive research around the best solutions of the population while considering the randomization that must be properly guided by Lévy flights for the exploration of new regions. Indeed, a standard CS extension is the addition of a method that handles the p_c fraction of smart cuckoos.

Algorithm 2: Improved Cuckoo Search

1: Objective function $f(x)$, $x = (x_1, \ldots, x_d)^T$
2: Generate the initial population of n nests $x_i (= 1, 2, \ldots, n)$
3: **while** (t <MaxCeneration) or (stop criteria) **do**
4:    **Lunch the search with a fraction (p_c) of smart cuckoos**
5:    Get a cuckoo randomly via Levy flights
6:    Evaluate its quality/fitness F_i
7:    Choose a nest among n (j) randomly
8:    **if** ($F_i < F_j$)(minimization) **then**
9:      Replace j by i
10:    **end if**
11:    a fraction (p_a) of worst nests is abandoned and new ones are built
12:    Hold the best solutions (or nests with high-quality solution)
13:    Classify the solutions and find the current best
14: **end while**
15: Post-process of results and visualization

The idea of improvement introduced in the CS algorithm is the search for solutions in regions specified by Lévy flights regardless of the best solution in the population. We also note that this improvement is a variant of the local search around a p_c fraction of the solutions. The principal disadvantage of local search is blocking in local optimum. This disadvantage is well dealt with in the improved CS which requires a displacement by region and not by solution, which remarkably reduces these blockages.

4.3 Discrete CS

In general, the process of adapting metaheuristics initially designed to solve continuous optimization problems to combinatorial optimization problems is not easy and not complete. Thus, the performance of the metaheuristic decreases compared to its version for continuous problems. On the other hand, it is important to keep adapted metaheuristics competitive in terms of result quality. The work in this chapter which focuses on designing an adaptation of CS to combinatorial optimization problems is described. The result is a discrete version of CS named "DCS".

To be simplified, the process of adaptation requires a decomposition of the problem into four elements: nest, egg, objective function, and search space.

4.3.1 Nest

In CS, a nest has the following characteristics:

- The number of nests is fixed. It represents the number of cuckoos that can pass to the next generation.

- A nest is a placeholder for an individual of the population.
- The number of nests is equal to the size of the population.
- An abandoned nest involves replacing an individual in the population with a new one.
- An egg in a nest represents a solution.

So, we can conclude from these characteristics that for an optimization problem, a nest appears as an individual of the population with its own solution.

4.3.2   Egg

A cuckoo can lay a single egg in a single nest, giving the eggs the following properties:

- An egg in a nest is a solution adopted by an individual of the population.
- A cuckoo's egg is a new candidate solution for a place in the population.
- An egg represents a next-generation cuckoo.

4.3.3   Objective Function

The objective function (test, fitness) is a function that, with each solution in the search space, associates a numerical value to describe its quality or fitness. For combinatorial optimization, this function will have discrete variables that composes a solution of problem. In CS, a solution is a cuckoo egg in the host nest. A nest of good egg quality will pass to the next generations, which means that the quality of a cuckoo egg is directly related to its ability to give a new cuckoo.

4.3.4   Search Space

In the case of two dimensions, the search space represents the potential positions of the nests. These positions are (x, y) coordinates in $\mathbb{R} \times \mathbb{R}$. To change a nest position, simply add (subtract) a real value to a coordinate (or two coordinates). As we notice, moving the nests is done without any real constraints. This is the case for the majority of continuous optimization problems, which can be considered as an advantage that avoids a set of technical obstacles such as

1. How to represent coordinates in a space of solutions of a combinatorial optimization problem?
2. Does this representation allow moving a solution to another neighbor?
3. How to define a neighborhood in this search space?

4. How to validate the length of the step that can be produced to pass from one solution to another near or far?

The search space in a combinatorial optimization problem can be seen as a "graph" where the vertices represent the solutions and the edges connect the neighboring pairs of solutions.

4.3.4.1 Coordinates

Coordinates are the elements that directly affect the quality of the solution through the objective function. This function must be well defined in order to simplify the calculation of the coordinates.

4.3.4.2 Displacement

In the combinatorial case, the coordinates of a solution in the search space are modified through the properties of the problem being dealt with. In general, the change of position in the combinatorial space is done by a change in the order of the elements of the solution, by a combination, a permutation, or a set of methods or operators called perturbations or moves.

4.3.4.3 Neighborhood

The concept of neighborhood requires that the solution close to a given solution must be generated by the smallest possible perturbation. This perturbation must make the minimum of changes on the current solution. It is therefore necessary to specify subsets of solutions named regions, according to the metrics of the search space. The notion of neighborhood has to reflect the neighborhood in terms of fitness. A neighborhood region is a region that contains solutions with small differences in their fitness.

4.3.4.4 Step

The step is the distance between two solutions. It is based on the topology of space and the concept of neighborhood. This model of CS classified the steps according to their length, which is the nature and/or the number of successive perturbation.

4.3.4.5   Lévy Flights

The aim of Lévy flights is to intensify the search around a solution, followed by occasionally big steps. And according to Yang and Deb (2009), the search for a new, better solution is more efficient via Lévy flights. So, to improve the quality of the search process, we will associate the length of the step with the value generated by the Lévy flights.

4.4   Conclusion

Swarm intelligence-based algorithms such as cuckoo search have a high performance to solve a wide range of nonlinear optimization problems, which explains their integration in different real-world application fields. This category of algorithms is designed to minimize the intelligence (or calculation provided) of the individual while maximizing the intelligence of the swarm in total. The minimization is accomplished with the reduction of the number of parameters, favoring more independence and abstraction with respect to the constraints of the problem dealt with. On the other hand, maximizing the intelligence of the swarm is achieved by better design of the solution space and a good balance of research between promising regions and globally for the exploration of new regions.

Thanks to the proposed enhancement and the discrete version of CS, DCS is designed in a way that maximizes the ability to adapt to a wide range of problems while keeping CS performance. The strong point of DCS is the ease of harmonizing three types of research through the best individual of the population, the fraction of bad solutions, and that of intelligent cuckoos. These three categories represent three independently operating strategies sharing information on promising regions and the position of the best solution.

The strengths of this approach can be summarized as follows:

- A high level of genericity, thanks to the reduced number of parameters;
- Controlled randomization via Lévy flights;
- A freedom of movement in the research space without being trapped by local optima, thanks to different search strategies.

The idea behind the design of DCS is to favor the consideration of the problem treated as a black box, with an input fed by disturbances or displacements (based on the definition of a solution) in the research space, and an output of the values of the objective function. In the rest of this book, all the three problems are solved by the improved version of CS. For this, DCS will designate improved DCS.

References

Bak P (1997) How nature works. Oxford University Press, Oxford

Brown CT, Liebovitch LS, Glendon R (2007) Lévy flights in dobe ju/?hoansi foraging patterns. Human Ecol 35(1):129–138

Fister I Jr, Yang X-S, Fister D, Fister I (2014) Cuckoo search: a brief literature review. In: Cuckoo search and firefly algorithm. Springer, pp 49–62

Gandomi AH, Talatahari S, Yang X-S, Deb S (2012) Design optimization of truss structures using cuckoo search algorithm. In: The structural design of tall and special buildings

Gandomi AH, Yang X-S, Alavi AH (2013) Cuckoo search algorithm: a metaheuristic approach to solve structural optimization problems. Eng Comput 29(1):17–35

Geem ZW, Kim JH et al (2001) A new heuristic optimization algorithm: harmony search. Simulation 76(2):60–68

Ouaarab A, Ahiod B, Yang X-S (2014) Discrete cuckoo search algorithm for the travelling salesman problem. Neural Comput Appl 24(7–8):1659–1669

Payne RB, Sorenson MD (2005) The cuckoos, vol 15. Oxford University Press, USA

Reynolds AM, Frye MA (2007) Free-flight odor tracking in drosophila is consistent with an optimal intermittent scale-free search. PLoS One 2(4):e354

Shlesinger MF, Zaslavsky GM, Frisch U (1995) Lévy flights and related topics in physics. In: Levy flights and related topics in physics, vol 450

Srivastava PR, Varshney A, Nama P, Yang X-S (2012) Software test effort estimation: a model based on cuckoo search. Int J Bio-Inspired Comput 4(5):278–285

Viswanathan G, Buldyrev S, Havlin S, Da Luz M, Raposo E, Stanley H (1999) Optimizing the success of random searches. Nature 401(6756):911–914

Yang C, Wang L, Liang W, Møller AP (2017) How cuckoos find and choose host nests for parasitism. Behav Ecol 28(3):859–865

Yang X-S, Deb S (2009) Cuckoo search via lévy flights. In: World congress on nature and biologically inspired computing, 2009. NaBIC 2009. IEEE, pp 210–214

Yang XS, Deb S (2010) Engineering optimisation by cuckoo search. Int J Math Model Numer Optim 1(4):330–343

Yang X-S, Deb S (2013) Multiobjective cuckoo search for design optimization. Comput Oper Res 40:1616–1624

Yang X-S, Deb S (2014) Cuckoo search: recent advances and applications. Neural Comput Appl 24(1):169–174

Yang X-S, Deb S, Karamanoglu M, He X (2012) Cuckoo search for business optimization applications

Part II
Application

Chapter 5
DCS Applications

Discrete versions of CS are applied to solve several combinatorial optimization problems (Yang and Deb 2014; Fister et al. 2014; Shehab et al. 2017). In all these applications, the real constraint is not enhancing the robustness of CS and solve efficiently these problems, but keeping its robustness in the adaptation and implementation phases for each specific problem such as optimal Distributed Generation (DG) allocation in a smart distribution grid (Buaklee and Hongesombut 2013), parallel machine scheduling (Guo et al. 2015), flow shop scheduling (Wang et al. 2017), optimal power flow (Mishra et al. 2015), graph coloring problem (Mahmoudi and Lotfi 2015), vehicle routing problem (Zheng et al. 2013), and reliability optimization problems (Valian et al. 2013). However, even if these phases are validated, generalizing DCS to solve all these problems and others is stilling an independent issue.

The idea behind adaptations developed in the contributions of this book is to allow one model of DCS to solve different problems with simplified and efficient procedures. This is what this book has mentioned in the previous chapters and will concrete in this part.

The CS adaptation strategy is implemented in this chapter. It presents in detail how the discrete version of CS is adapted to search and generate solutions in the combinatorial space.

It begins with a discussion of the DCS application on the three predefined COPs, where a description is about how the reinterpretation of CS strategies is implemented. These strategies are the key to the transition from continuous to combinatorial space. Therefore, while taking these three problems as typical examples of combinatorial optimization problems, the following important concepts are discussed before its implementation: position (or solution presentation), displacement, distance that is related to the notion of neighborhood, and objective function. A detailed description of each one of these strategies is given, and their performances relative to different COPs are validated by the numerical test results on different groups of benchmark instances.

© Springer Nature Singapore Pte Ltd. 2020
A. Ouaarab, *Discrete Cuckoo Search for Combinatorial Optimization*,
Springer Tracts in Nature-Inspired Computing,
https://doi.org/10.1007/978-981-15-3836-0_5

5.1   Main Functions

5.1.1   Get Cuckoo

Get Cuckoo (getCS) is the main function of CS algorithm. It searches via Lévy flights for a new cuckoo around the best solution of the population. When the new solution is found, its fitness is compared to an individual chosen randomly from the population. If the new cuckoo is better it will take its place in the population, else the individual keeps its place. This procedure is implemented in Algorithm 3.

Algorithm 3: GetCS function

Input: Sol
cuckoo = Sol;
/* Sol is the best solution of the population   */
Choose an individual *ind* randomly in the population;
if *Lévy(λ)* < 0.5 **then**
  | cuckoo=localSearch(smallStep(Sol));
else
  | cuckoo=localSearch(bigStep(Sol));
end
if *Fitness(cuckoo)* < *Fitness(Sol)* **then**
  | replace *ind* by cuckoo;
end

5.1.2   Smart Cuckoo

Smart cuckoos (smartCS) are proposed by the improved version of CS. It is based on a portion of cuckoos that improve themselves by searching around the best solution of the population. This procedure works differently compared to getCS method. In getCS, the improvement is on a random individual in the population. However, in smartCS of Algorithm 4, the improvement is on each individual of the portion.

Algorithm 4: SmartCS function

Input: Sol,pc
csSol = Sol;
/* Sol is the best solution of the population   */
for $i \leftarrow 1$ **to** $Dim \times pc$ **do**
  if $Lévy(\lambda) < 0.5$ **then**
  | csSol=localSearch(smallStep(Sol));
  else
  | csSol=localSearch(bigStep(Sol));
  end
  if $Fitness(csSol) < Fitness(Sol)$ **then**
  | replace Sol by csSol;
  end
end

5.1.3 Worst Cuckoo

As described by CS, cuckoos that are not able to find good nests their eggs well be discovered by the host bird and will not pass to the next generation. For that, DCS has proposed worst cuckoo mechanism, implemented by worstCS function in Algorithm 5. It replaces a portion p_a of worst cuckoos from the population by new ones far from the best nest.

Algorithm 5: WorstCS function

Input: Sol,p_a
worstSol = Sol;
/* Sol is the best solution of the population   */
for $i \leftarrow n \times (1 - p_a)$ **to** $Dim - 1$ **do**
  worstSol=localSearch(bigStep(Sol));
  if $Fitness(population(i)) < Fitness(worsSol)$ **then**
  | replace population(i) by worstSol;
  end
end

5.2 TSP Adaptation

DCS, while solving TSP, begins with rebuilding the CS population by introducing a new category of cuckoos with the aim of effectively improving its research. One of the goals of the CS extension is to study the integration of the basic advantages of

CS into its discrete version. The procedure for adapting CS to TSP focuses mainly on a reinterpretation of the terminology used in the standard CS. CS and its source of inspiration are structured around five main elements: egg, nest, objective function, research space, and Lévy flights. In this section, these notions will be implemented to be applied for TSP instances.

5.2.1 Egg and Nest

In the case where a solution is the equivalent of an egg, a location/position is reserved for each individual in the population. To be adapted to TSP, an egg is the equivalent of a Hamiltonian cycle (Fig. 5.1). Here, we neglect the need to take the departure city for all the circuits and also the direction of the tour taken by the traveler (symmetric problem).

In DCS, the Hamiltonian cycle is implemented by a series of integer numbers (cities) placed regarding the salesman visiting order. The generation of an egg (solution) is performed randomly based on the "Random solution" function in Algorithm 6.

Algorithm 6: Random solution

Result: Sol

Create an empty solution Sol;

for $i \leftarrow 0$ **to** $Dim - 1$ **do**

| Sol(i) $\leftarrow$ new random city;

end

```
/* Add the departure city in the end                    */
```

Sol(Dim) = Sol(0);

return *Sol*

Additionally, for a TSP solution, we can assume that a nest is seen as an individual in the population with its own Hamiltonian cycle. Indeed, a nest can have several

Fig. 5.1 Hamiltonian cycle: represents one egg

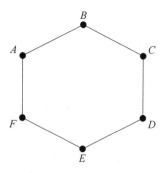

eggs in alternative extensions. In the case of DCS, each nest contains only one egg, and the population is constructed as shown in Algorithm 7.

Algorithm 7: Population construction

Result: Pop
for $i \leftarrow 0$ **to** $n - 1$ **do**
  | Pop(i) = random solution;
end
return *Pop*

5.2.2 Objective Function

In order to evaluate the fitness of a given solution, TSP associates with each solution a quality value, the length of the Hamiltonian cycle. The best solution is the one with the shortest Hamiltonian cycle. This quality value (to be minimized) is calculated by the objective function as presented in Algorithm 8.

Algorithm 8: Fitness

Input: Sol
Result: Fit
Fit = 0;
for $i \leftarrow 0$ **to** $Dim - 1$ **do**
  | Fit + = Distance between city $Sol(i)$ and $Sol(i + 1)$;
end
return *Fit*

5.2.3 Search Space

In the two-dimensional spaces, CS represents the positions of the potential nests by the coordinates $(x, y) \in \mathbb{R} \times \mathbb{R}$. If the space is continuous, changing the nest position is done by modifying only the current value of each of its coordinates, which is considered as an advantage for continuous optimization. This advantage can avoid several technical obstacles like the representation of coordinates in the TSP solution space, especially during the displacement of a solution toward another neighbor.

Each city in TSP is represented by its real coordinates $(x, y) \in \mathbb{R} \times \mathbb{R}$. These coordinates are used to generate the distance matrix that contains distances between cities. To solve TSP, DCS replaces the coordinates (x, y) by the visiting order of the city.

Fig. 5.2 TSP solution space
of an instance of five cities

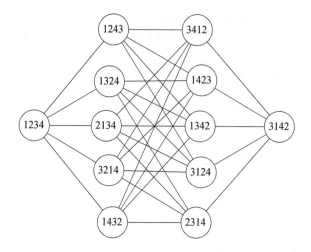

As shown in Fig. 5.2, the search space of TSP is represented by a convex graph where vertices are the solutions and arcs are 2-opt move. A tour is a potential solution with $n = 5$ cities (cities are identified by the integers from "0" to "4", so a solution is a sequence/permutation of these integers. In the graph, the starting city identified by "0" is fixed; this is why it is not shown). These solutions are positioned in the space according to the order of their cities. In this example, we have 12 distinct solutions in the search space (structured using the 2-opt operator), and each solution is directly connected with $n(n - 3)/2$ neighbors.

5.2.3.1   Moving in the Search Space

Because the coordinates of the cities are fixed, the movements are based on the city order of visit. Regarding this constraint, there are several moves, operators, or perturbations that generate a new solution from an existing solution.

The adaptation of CS to TSP (Discrete CS) perturbs a solution by changing the visiting city order. The proposed moves in this version of DCS for TSP are 2-opt (Croes 1958) and double-bridge moves. 2-opt, as shown in Fig. 5.3, deletes two nonadjacent arcs of the tour and reconnects the two created paths. Double bridge cuts four arcs and introduces four new arcs as shown in Fig. 5.4.

The displacement unit is used to calculate the distance between two solutions. It is related to the topology of space and the notion of neighborhood. The length of the step is proportional to the number of successive 2-opt moves. A big step is represented by a double-bridge move.

Fig. 5.3 2-opt move. **a** Initial tour. **b** New generated tour by 2-opt

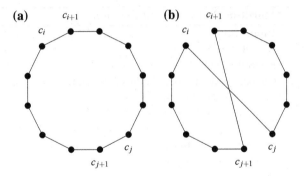

Fig. 5.4 Double-bridge move. **a** Initial tour. **b** New generated tour by double-bridge move

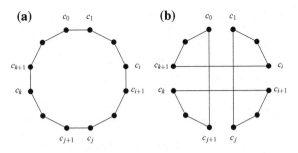

Moving from one solution to another in the search space is done by DCS in three different ways: (1) A small step performed by 2-opt move; (2) A bigger step done by a succession of small steps (2-opt moves) in a restricted area around the current solution. So, these steps lead to relatively distant solutions; (3) A big jump where the move is from the current region to another very distant region. This jump is performed by the double-bridge operator. The corresponding algorithms 9 and 10 show how DCS proposes its two principal moves.

Algorithm 9: 2-opt move on arcs (c_i, c_{i+1}) and (c_{j-1}, c_j).

Input: Sol, c_i, c_j
Result: newSol
newSol = Sol;
Inverse the order of cities between c_i and c_j of newSol;
return *newSol*

Algorithm 10: Double-bridge move

Input: Sol
Result: newSol
newSol = Sol;
Choose random ordered cities c_i, c_j and c_k;
for *iter* ← c_1 **to** c_i **do**
| Add Sol(iter) to newSol;
end
for *iter* ← c_{k+1} **to** c_0 **do**
| Add Sol(iter) to newSol;
end
for *iter* ← c_{j+1} **to** c_k **do**
| Add Sol(iter) to newSol;
end
for *iter* ← c_{i+1} **to** c_j **do**
| Add Sol(iter) to newSol;
end
Add Sol(Dim) to newSol;

The newSol created by double-bridge move is the result of removing the nonadjacent edges (c_0, c_1), (c_i, c_{i+1}), (c_j, c_{j+1}), and (c_k, c_{k+1}) which are replaced by the edges (c_0, c_{j+1}), (c_i, c_{k+1}), (c_j, c_1), and (c_k, c_{i+1}), respectively.

5.2.3.2   Neighborhood

In continuous problems, the definition of neighborhood is clear and simple relatively combinatorial optimization problems, where this notion of requires that the neighborhood of a given solution is generated by the smallest step perturbation. This perturbation must cause the minimum possible change on the solution, which led us to the 2-opt move because, for a new solution, the minimum number of non-contiguous arcs we can remove is two. So, 2-opt is a good candidate for this type of perturbations. For this reason, DCS performed 2-opt to search locally in the neighborhood of the current solution and proposed a descent-based local search method as developed in Algorithm 11.

Algorithm 11: Local search

Input: Sol
Result: neighbor
neighbor = Sol;
Choose random ordered cities c_i, c_j and c_k;
for $i \leftarrow 0$ **to** $Dim - 2$ **do**
    for *Every move with j and i* **do**
        if *The move improves neighbor* **then**
             neighbor=2-optMove(neighbor);
        end
    end
end

Algorithm 11 searches in the neighborhood of a current solution. It makes successive first improving steps. Each step is a 2-opt move, and each step improves the current solution until no improvement is possible. Then, a local optimum is returned.

5.2.3.3 Lévy Flights

Lévy flights have a characteristic, intensified search started around a solution, followed by a big step in the long term. According to Yang and Deb (2009), in some optimization problems, the search for a new, much better solution is more efficient via Lévy flights. In order to improve search quality, we will associate the length of the step with the value generated by Lévy flights as presented in the standard CS version.

The solution space must support the notion of step. As predefined in this section, the selected step unit for moving from one solution to another is 2-opt move. So we have the choice between a small step, a number k of steps, and a big step. To facilitate the control of these steps via Lévy flights, we associate them with an interval between 0 and 1. Indeed, according to the value returned by Lévy flights in this interval, we can choose the corresponding step length.

Algorithm 12: Lévy flights

Input: λ, $\lambda \in]1, 3]$
Result: Lévy
Lévy = $s^{-\lambda}$;

Based on the values returned by Lévy function in Algorithm 12, a generalized example can explain how Lévy works with DCS. If the value of Lévy is in,

1. $[0, i[$ we have a step of one 2-opt move,
2. $[(k-1) \times i, k \times i[$ a step by k moves, and
3. $[k \times i, 1[$ we take a big step by double-bridge move.

The value of i in this procedure is $i = (1/(n+1))$, where n is the maximum number of steps and k is $\{2, \ldots, n\}$.

Suppose that $n = 4$, so $i = 0.2$, then our interval is divided into five parts.

Lévy in $[0, i[\longrightarrow [0, 0.2[\longrightarrow$ one step by 2-opt move.

"   " $[i, i \times 2[\longrightarrow [0.2, 0.4[\longrightarrow$ two steps by 2-opt.

"   " $[i \times 2, i \times 3[\longrightarrow [0.4, 0.6[\longrightarrow$ three steps by 2-opt.

"   " $[i \times 3, i \times 4[\longrightarrow [0.6, 0.8[\longrightarrow$ four steps by 2-opt.

"   " $[i \times 4, 1[\longrightarrow [0.8, 1[\longrightarrow$ one step by double-bridge.

5.2.4 Experimental Results

To evaluate the performance of DCS, validation tests are performed against a subset of benchmark TSP instances. Each instance is read from a ".tsp" file, taking the name of the instance, by "Tsp.java" class implemented in Sect. B.1.1 in Appendix B. This class get information such as its name, type, edge weight type, dimension, and coordinates of the cities. It generates the distance matrix and implements the fitness function. More detailed description about the components of the TSP file are in Appendix A.1.

DCS is tested by a group of benchmark TSP instances taken from the publicly available Traveling Salesman Problem Library (TSPLIB) (Reinelt 1991). Most of the instances included in TSPLIB have already been solved in the literature, and their found optimal values can be used to compare the algorithms. Forty-one instances are considered with a size between 51 and 1379 cities. In Reinelt (1991), all these TSP instances belong to the type of Euclidean instances. A TSP instance is made up of cities with their coordinates. The numeric values in the instance label represents the number of cities provided, for example, the instance name "eil51" indicates that it has 51 cities.

DCS is implemented using Java language under the 32bit Vista OS. The tests were conducted in an Intel(R) Core 2 Duo 2.00 GHz CPU, and 3G RAM. The parameter values of the proposed algorithm are selected based on a set of preliminary tests. The parameters selected for DCS are those that give the best values of the presented results, concerning the quality of the solution or the calculation time necessary to reach this solution. The parameters set for the tests are shown in Table 5.1. In each case study, 30 executions independent of the algorithm with the selected parameters are reported. More test results are presented in Ouaarab et al. (2014).

Table 5.2 summarizes the experimental results, where its columns are as follows:

- "instance": the instance name,
- "opt": the optimal solution length taken from TSPLIB,
- "best": the best solution length found by each algorithm,

Table 5.1 Parameters of both algorithms DCS, standard and improved

Parameter	Value	Description
n	20	Population size
p_a	0.2	Portion of worst solutions
p_c	0.6	Portion of smart cuckoos (Just, in the case of improved DCS)
$MaxGeneration$	500	Maximum number of iterations

- **"average"**: average length of the solution of 30 executions independent of each algorithm,
- **"worst"**: length of the worst solution found by each algorithm,
- **"SD"**: standard deviation,
- **"PDav(%)"**: percentage of the average length deviation from the length of the optimal solution during 30 runs,
- **"PDbest(%)"**: deviation percentage of the length of the best solution from the length of the optimal solution over 30 runs,
- **"$C_{1\%}/C_{opt}$"**: the number of solutions that are within 1% optimality (over 30 runs)/the number of the optimal solutions, and
- **"time"**: average time in seconds over 30 runs.

The percentage deviation of the current solution from the best known solution is given by the formula in Eq. 5.1:

$$PD_{solution}(\%) = \frac{\text{solution length} - \text{best known solution length}}{\text{best known solution length}} \times 100 \qquad (5.1)$$

Table 5.2 and Fig. 5.5 show the experimental results of DCS on 41 TSPLIB euclidean symmetric instances. Considering $PDbest(\%)$, we can say that 90.24% value of $PDbest(\%)$ is less than 0.5%, which explains that the best solution found, of 30 executions, is close to less than 0.5% of the best known solution, while the 0.00 shown in bold in the $PDav(\%)$ column indicates that all solutions found during 30 runs have the same cost as the best solution.

These results explain in fact the robustness of CS: a good balance between intensification and diversification, intelligent use of Lévy flights, and the reduced number of control parameters. It is further explained by the improved structure of the population which encompasses a variety of cuckoo-based search methods. One of the benefits of the improved DCS is the relative independence of cuckoos when looking for a new solution around the best position. So, it is more likely to find good solutions in unexplored areas, starting from the best solution with relative independence from it.

Table 5.2 Experimental results of DCS algorithm for 41 TSP instances

Instance	Opt	Best	Worst	Average	SD	PDav(%)	PDbest(%)	$C_{1\%}/C_{opt}$	Time
eil51	426	**426**	**426**	**426**	**0.00**	**0.00**	**0.00**	**30/30**	1.16
berlin52	7542	**7542**	**7542**	**7542**	**0.00**	**0.00**	**0.00**	**30/30**	0.09
st70	675	**675**	**675**	**675**	**0.00**	**0.00**	**0.00**	**30/30**	1.56
pr76	108159	**108159**	**108159**	**108159**	**0.00**	**0.00**	**0.00**	**30/30**	4.73
eil76	538	**538**	539	538.03	0.17	0.00	**0.00**	30/29	6.54
kroA100	21282	**21282**	**21282**	**21282**	0.00	**0.00**	**0.00**	**30/30**	2.70
kroB100	22141	**22141**	22157	22141.53	2.87	**0.00**	**0.00**	30/29	8.74
kroC100	20749	**20749**	**20749**	**20749**	**0.00**	**0.00**	**0.00**	**30/30**	3.36
kroD100	21294	**21294**	21389	21304.33	21.79	0.04	**0.00**	30/19	8.35
kroE100	22068	**22068**	22121	2281.26	18.50	0.06	**0.00**	30/18	14.18
eil101	629	**629**	633	630.43	1.14	0.22	**0.00**	30/6	18.74
lin105	14379	**14379**	**14379**	**14379**	**0.00**	**0.00**	**0.00**	**30/30**	5.01
pr107	44303	**44303**	44358	44307.06	12.90	0.00	**0.00**	30/27	12.89
pr124	59030	**59030**	**59030**	**59030**	**0.00**	**0.00**	**0.00**	**30/30**	3.36
bier127	118282	**118282**	118730	118359.63	12.73	0.06	**0.00**	30/18	25.50
ch130	6110	**6110**	6174	6135.96	21.24	0.42	**0.00**	28/7	23.12
pr136	96772	96790	97318	97009.26	134.43	0.24	0.01	30/0	35.82
pr144	58537	**58537**	**58537**	**58537**	**0.00**	**0.00**	**0.00**	**30/30**	2.96
ch150	6528	**6528**	6611	6549.9	20.51	0.33	**0.00**	29/10	27.74
kroA150	26524	**26524**	26767	26569.26	56.26	0.17	**0.00**	30/7	31.23
kroB150	26130	**26130**	26229	26159.3	34.72	0.11	**0.00**	30/5	33.01
pr152	73682	**73682**	**73682**	**73682**	**0.00**	**0.00**	**0.00**	**30/30**	14.86
rat195	2323	2324	2357	2341.86	8.49	0.81	0.04	20/0	57.25
d198	15780	15781	15852	15807.66	17.02	0.17	0.00	30/0	59.95
kroA200	29368	29382	29886	29446.66	95.68	0.26	0.04	29/0	62.08
kroB200	29437	29448	29819	29542.49	92.17	0.29	0.03	28/0	64.06
ts225	126643	**126643**	126810	126659.23	44.59	0.01	**0.00**	30/26	47.51
tsp225	3916	**3916**	3997	3958.76	20.73	1.09	**0.00**	9/1	76.16
pr226	80369	**80369**	80620	80386.66	60.31	0.02	**0.00**	30/19	50.00
gil262	2378	2382	2418	2394.5	9.56	0.68	0.16	22/0	102.39
pr264	49135	**49135**	49692	49257.5	159.98	0.24	**0.00**	28/13	82.93
a280	2579	**2579**	2623	2592.33	11.86	0.51	**0.00**	25/4	115.57
pr299	48191	48207	48753	48470.53	131.79	0.58	0.03	27/0	138.20
lin318	42029	42125	42890	42434.73	185.43	0.96	0.22	15/0	156.17
rd400	15281	15447	15704	15533.73	60.56	1.65	1.08	0/0	264.94
fl417	11861	11873	11975	11910.53	20.45	0.41	0.10	30/0	274.59
pr439	107217	107447	109013	107960.5	438.15	0.69	0.21	22/0	308.75
rat575	6773	6896	7039	6956.73	35.74	2.71	1.81	0/0	506.67
rat783	8806	9043	9171	9109.26	38.09	3.44	2.69	0/0	968.66
pr1002	259045	266508	271660	268630.03	1126.86	3.70	2.88	0/0	1662.61
nrw1379	56638	58951	59837	59349.53	213.89	4.78	4.08	0/0	3160.47

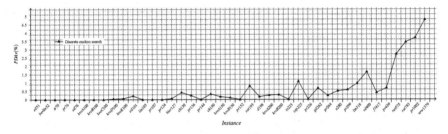

Fig. 5.5 PDav(%) (out of 30 runs) for 14 instances of TSPLIB

5.3 JSSP Adaptations

The goal of scheduling problems is to find a feasible order in an infinite or finite countable set that optimizes the proposed objective function. This category of problems is generally used in a multitude of manufacturing or service industries. One of the most well-known scheduling problems is the JSSP (Applegate and Cook 1991; Manne 1960).

When resolving the JSSP, DCS adopts its own procedure without using additional enhancements to achieve good results. In DCS, Lévy flights have control over all movements in the local and global searches. For that, this section shows how to present a solution in space and how to move from the current solution to another using Lévy flights. The objective function that is related to the neighborhood local search will be presented and implemented.

5.3.1 JSSP Solution

A nest with an egg represents an individual, with a solution, in the population. Therefore, DCS considers scheduling as a solution in the search space. The illustration of a JSSP instance uses the data example of Table 5.3 that is composed of 3 jobs and 4 machines, and each job operation is associated with its corresponding machine by operation processing time. A solution of this JSSP instance is a permutation of operations, which can be generated randomly, in the case of an initial solution as shown in Fig. 5.6.

Algorithm 13: JSSP solution

Input: Number of jobs j and machines m
Result: Sol
for $i \leftarrow 0$ **to** $j \times m$ **do**
| Sol=$i \bmod j$;
end
for $i \leftarrow 0$ **to** $j \times m \div 4$ **do**
| Swap positions in Sol randomly;
end

First, a series of random integers is generated. For this, DCS implements a solution generator algorithm 13. On each number in this series, we apply the operation "$(i \bmod j) + 1$", where i is a number in the series and j is the number of jobs, which is 3 in the example instance. So the result is a job permutation that can be interpreted as a permutation of operations (Fig. 5.6).

In DCS, a scheduling is an egg in the nest, which is able to switch to the next generation of cuckoos through hatching. Therefore, a schedule with a good completion time is selected to produce the next generation of schedules. Based on Table 5.3 and the solution illustrated in Fig. 5.6, the Gantt diagram (Fig. 5.7) is built to represent this solution (schedule) and the completion time of its job operations in each machine.

Table 5.3 3×4 JSSP instance

Jobs	Machine sequence	Processing times			
1	1 2 3 4	$p_{11} = 10$	$p_{12} = 8$	$p_{13} = 4$	$p_{14} = 2$
2	2 1 4 3	$p_{21} = 8$	$p_{22} = 3$	$p_{23} = 5$	$p_{24} = 6$
3	1 2 4 3	$p_{31} = 4$	$p_{32} = 7$	$p_{33} = 3$	$p_{34} = 5$

Random integer series	4	10	6	1	8	3	5	2	9	7	11	12
Job permutation	2	2	1	2	3	1	3	3	1	2	3	1
Operations' permutation	O_{21}	O_{22}	O_{11}	O_{23}	O_{31}	O_{12}	O_{32}	O_{33}	O_{13}	O_{24}	O_{34}	O_{14}

Fig. 5.6 Procedure for random generation of an initial solution

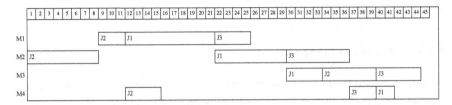

Fig. 5.7 Gantt chart of the solution shown in Fig. 5.6

The initial population is built simply by random solutions as described in Algorithm 14.

Algorithm 14: JSSP population

Result: *Population*
for $i \leftarrow 0$ **to** *population size* **do**
  Generate a random solution *Sol*;
  Add *Sol* to *Population* ;
end

5.3.2 Objective Function

The objective function to be minimized is a function that returns the maximum completion time (makespan) of a schedule. The optimum is a schedule that has the minimum makespan among all feasible schedules. In DCS, the fitness of each cuckoo (schedule) is associated with its makespan as calculated in Algorithm 15.

Algorithm 15: JSSP fitness

Input: Sol
Result: Makespan
Machine for machines max time update;
Job for jobs max time update;
Jind for the index of the current machine operation;
```
/* Machine, Jobs and Jind are initialized by zeros    */
/* N and M are numbers of jobs and machines           */
```
for $i \leftarrow 0$ **to** $N \times M$ **do**
  Get the current job j from $Sol(i)$;
  Get the current machine operation m from $Jind(j)$;
  Get the max(Machine(current machine,Job(current job));
  Update *Machine*, *Job* and *Jind*;
end
Get max of machines' completion time;

5.3.3   Search Space

5.3.3.1   Moving in the Space

In order to move from one solution to another in the combinatorial search space, we can alternate between the notion of step length and topology. The step length involves defining the distances between the solutions in a given search space, and then we will be able to describe a small or big step. In the case of the topology, we move from one solution of the current topology to another in the new topology. This involves the introduction of different types of moves. So, we are talking about a step toward a different topology.

In DCS application on JSSP instances, three different moves, controlled by Lévy flights, are proposed. Considering the following figures, the first presents insertion operator (Fig. 5.8) that removes the integer from the specified position "2" and inserts it into the second specified position "6". The second operator used is swap operator (Fig. 5.9) which exchanges the positions of two integers, respectively, located in the positions "2" and "6". The last one is the inversion operator (Fig. 5.10) that reverses the order of integers between two positions, "2" and "6" in the example.

These moves are associated with random values generated by Lévy flights distribution. Indeed, the search in the JSSP solution space is balanced between restricted regions and wide new topologies.

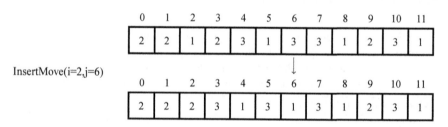

Fig. 5.8 Displacement by insertion operator

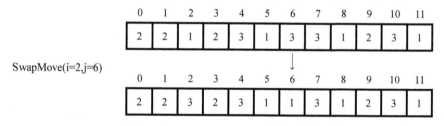

Fig. 5.9 Displacement by swap operator

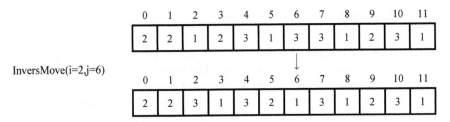

InversMove(i=2,j=6)

Fig. 5.10 Displacement by inversion operator

5.3.3.2 Neighborhood

The nearest schedule to the current solution is proposed by DCS as the result of swap move. It is then applied to search in the neighborhood of a solution. Therefore, local search is a successive improving swap moves until the local optimum is reached.

Algorithm 16: Local Search

Input: Sol
Result: localOpt
localOpt=Sol;
for $i \leftarrow 0$ **to** n, $n < Dim$ **do**
    Get a couple i and j randomly;
    /* Swap move between i and j   */
    Sol=Swap(localOpt,i,j);
    if *Sol is better than localOpt* **then**
      | localOpt=Sol;
    end
end

5.3.3.3 Lévy Flights

An example of using values generated by Lévy flights, where three intervals are proposed, is shown as follows:

- Lévy$(\lambda)\in [0, 0.3] \longrightarrow$ Swap move;
- Lévy$(\lambda)\in [0.3, 0.6] \longrightarrow$ Insert move;
- Lévy$(\lambda)\in [0.6, 1] \longrightarrow$ Inverse move.

Among the three moves of this example, the most frequent is Swap that moves to the nearest solution. On the other side, the least frequent is Inverse that jumps far from the current solution.

5.3.4 Experimental Results

DCS reads JSSP files (.jsp) in order to get the instance data. It reads from the second line dimensions to create "Machines" and "Durations" matrices. "Machines" and "Durations" will contain odd and even columns of the table.

Before solving JSSP instances, some preliminary trials are launched, and the following parameter settings are selected as shown in Table 5.4.

In each case study, 10 DCS independent executions are performed. Table 5.5 summarizes the experimental results, where the first column shows the name of the instance and the best known solution in parentheses, the "best" column shows the solution of the best completion time found by DCS, the "average" column designates the average completion time of DCS-independent executions, and the "PDav(%)"

Table 5.4 Parameter values of DCS for JSSP

Parameter	Value	Description
n	30	Population size
P_a	0.2	Potion of worst solutions
p_c	0.6	Portion smart cuckoos
$MaxGeneration$	300	Maximum number of iterations
α	0.01	Step size
λ	1	Index

Table 5.5 Test results of applying DCS on 16 JSSP instances

Instance (opt)	Best	Average	$PDav(\%)$
Abz5 (1234)	1239	1239.6	0.00
Abz6 (943)	943	946.4	0.00
Ft06 (55)	**55**	**55**	**0.00**
Ft10 (930)	945	966.8	0.03
Ft20 (1165)	1173	1178.8	0.01
La01 (666)	**666**	**666**	**0.00**
La02 (655)	**655**	**655**	**0.00**
La03 (597)	604	607	0.01
La04 (590)	**590**	**590**	**0.00**
La06 (926)	**926**	**926**	**0.00**
La11 (1222)	**1222**	**1222**	**0.00**
La16 (945)	946	967.7	0.02
La21 (1046)	1055	1070.1	0.02
La26 (1218)	**1218**	**1218**	**0.00**
La31 (1784)	**1784**	**1784**	**0.00**
La36 (1268)	1297	1313.6	0.03

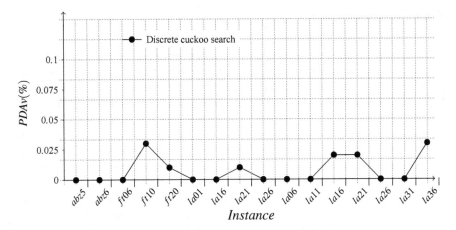

Fig. 5.11 PDav(%) of 16 JSSP instances solved by DCS

column designates the percent deviation of the average of the optimal solution completion times for 10 executions. All values in bold mean that the found solutions have the same completion time on all executions.

As shown in Table 5.5, tests are performed on a set of benchmark JSSP instances. Hence, DCS is more suitable to be adapted to solve the JSSP and provide good results (Fig. 5.11), which validate the well-controlled balance between an exploration of space toward different regions and exploitation of promising areas, thanks to Lévy flights, and also by adding enhancement guided by smart cuckoos category.

5.4 QAP Adaptations

The quadratic assignment problem is one of the typical problems of combinatorial optimization (Taillard 1991). It is NP-hard according to Sahni and Gonzalez Sahni and Gonzalez (1976) during its original development by Koopmans and Beckmann (1957). Through this combinatorial optimization problem, we will have the possibility to treat a multitude of real problems, in different domains of application adopting the QAP as their model of abstraction. These applications include manufacturing plant design and planning, vehicle routing issues, network data, layout planning, and large-scale manufacturing system design, among others (Davendra and Zelinka (2012)).

The key strategy of DCS is the isolation of the useful components of the problem and associating them with its terminology. Indeed, before launching the QAP resolution, each component is defined to make it more controllable by DCS. Following the design of the QAP, we have the ability to handle three main concepts: QAP solution, objective function, and displacement steps (moves).

5.4.1 QAP Solution

As mentioned by its description, QAP is a combinatorial optimization problem that offers a set of solutions scattered in the search space. These solutions simply represent all possible potential assignments of the facilities to the locations. DCS considers a QAP solution S as a permutation, written under the formula: $S = \{\phi(1), \phi(2), \ldots, \phi(N)\}$, where N designates the dimension of the instance of the problem, which is also the number of facilities and locations, and $\phi(i) = k$ means that the facility i is assigned to the location k. The solution feasibility constraints in QAP are not formulated because the only requirement is that at each location we assign a unique facility and vice versa.

A QAP solution is relatively simple to represent. It takes the form of a series of integers similar in size to the number of facilities in the instance. Integers are the facility indexes that are ordered according to their location assignments, as shown in Fig. 5.12.

Solutions like the one in Fig. 5.12 are built in the following function (Algorithm 17). After building solutions, they are placed in the initial population as shown in Algorithm 18.

Algorithm 17: QAP Solution

Result: Sol
for $i \leftarrow 0$ **to** *Dim* **do**
| Add i to Sol;
end
for $i \leftarrow 0$ **to** *Dim* $\div$ 4 **do**
| swap i and a random position in Sol;
end

Algorithm 18: QAP population

Result: Population
for $i \leftarrow 0$ **to** *population size* **do**
| Generate a random solution Sol;
| Add Sol to Population;
end

Fig. 5.12 QAP solution

Facility assignments

| 5 | 1 | 4 | 6 | 3 | 8 | 7 | 2 |

5.4.2   Objective Function

After defining a QAP solution, the second phase of the adaptation procedure is to have an idea about the difference in qualities between two solutions in the search space. The objective function, as its nomination indicates, is a value associated with each solution in order to reflect its quality according to a goal predefined by the problem. This goal is for QAP, finding the best assignment in the search space, guided by the objective function. Achieving this best solution is due to minimizing the generated value of the function implemented in Algorithm 19.

Algorithm 19: QAP Objective function

Input: Sol
Result: Fitness
for $i \leftarrow 0$ **to** *Sol size* **do**
    for $j \leftarrow 0$ **to** *Sol size* **do**
        `/* As in equation 5.2)`   `*/`
        Fitness=Fitness+$a_{ij}b_\phi(i)\phi(j)$;
    end
end

The objective function of QAP leads, therefore, to the search for regions containing a permutation $\phi = (\phi(1), \phi(2), \ldots, \phi(N))$ which minimizes the value of the following equation:

$$\sum_{i=1}^{N}\sum_{j=1}^{N} a_{ij}b_\phi(i)\phi(j), \qquad (5.2)$$

where a_{ij} is the distance between the locations i and j, and the flow of the facility i to the facility j takes the value $b_{\phi(i)\phi(j)}$. The goal is to minimize the sum of distance×flow (Taillard 1991).

5.4.3   Search Space

5.4.3.1   Step and Displacement

In a general context, a displacement in the combinatorial search space is the result of any perturbation applied on the current solution. Especially in our approach, for the representation of displacement steps, DCS adopted two operators: swap as a small step named **smallStep** in Algorithm 20 and the second is a big step named **bigStep** in Algorithm 21 and performed by successive swaps.

Algorithm 20: QAP small step move
___
Input: Sol
Result: SmallStep
Choose two random positions p_1 and p_2 in Sol;
SmallStep=Swap p_1 and p_2 positions;

Algorithm 21: QAP big step move
___
Input: Sol
Result: bigStep
Choose k random positions p in Sol;
for $i \leftarrow 0$ **to** k **do**
  | bigStep=swap(Sol,i,p_i) ;
end

Considering the current location ϕ, we can move to a neighbor location π by applying a swap between facilities r and s:

$$\pi(k) = \phi(k), \forall k \neq r, s \tag{5.3}$$

$$\pi(r) = \phi(s), \pi(s) = \phi(r). \tag{5.4}$$

The tested instances in this approach have all symmetric matrices with a zero diagonal. So, the cost $\delta(\phi, r, s)$ of a swap displacement between facilities r and s is given by

$$\delta(\phi, r, s) = \sum_{i=1}^{N} \sum_{j=1}^{N} (a_{ij} b_{\phi(i)\phi(j)} - a_{ij} b_{\pi(i)\pi(j)}) \tag{5.5}$$

$$= 2. \sum_{k \neq r,s} (a_{sk} - a_{rk})(b_{\phi(s)\phi(k)} - b_{\phi(r)\phi(k)}). \tag{5.6}$$

5.4.3.2 Neighborhood

In order to minimize the calculation time, DCS tries to search efficiently in the solution space. It performs an estimation strategy before making a move. This way, only improving moves will be performed, which restricts positively the choice to move to a new and better solution. Therefore, local search tries to find the local optimum by making estimates before each move. QAP local search procedure is presented in Algorithm 22.

Algorithm 22: Local Search

Input: Sol
Result: localOpt
localOpt=Sol;
for $i \leftarrow 0$ **to** $n - 2$ **do**
    for $j \leftarrow i + 2$ **to** $n - 1$ **do**
        Sol=Swap(localOpt,i,j);
        if *Sol is better than localOpt* **then**
            localOpt=Sol;
        end
    end
end

5.4.3.3 Lévy Flights

Starting from their description in DCS, Lévy flights require more details about the definition of a small and a big step to design a significant move from the current solution to a new one. A small step is an application of a swap, one time, and a big step is represented by a sequence of swaps applied on the same solution. The number of swaps (step length) is associated with the values generated by Lévy flights. It is then simply illustrated as follows:

- Lévy(λ)$\in [0, 0.5[\longrightarrow$ one swap move;
- Lévy(λ)$\in [0.5, 1] \longrightarrow k$ (to be defined) swap moves.

5.4.4 Experimental Results

DCS reads in QAP input file (of Appendix A, Section A.3) the instance dimension, the flow, and distance matrices. By the class "QAP.java" implemented in Sect. B.3.1 in Appendix B, it creates an object QAP that handles this instance data and calculates its fitness.

Table 5.6 shows the numerical results reported from the implementation of DCS to solve some reference instances (Taillard) of the QAP taken from the library QAPLIB (Burkard et al. 1997). We note that for each instance, 15 independent executions were performed. In this direct application of DCS to QAP, we applied a single type of moves (the swap), which allowed us to stay in the same topology of the QAP search space. We have not proposed any advanced local search method, or solution perturbation by complex techniques in order to perform a free and inde-

Table 5.6 Numeric results of the DCS application on QAP benchmark instances

Instance	Bkv	Best	Average	Worst	PDav(%)
tai12a	224416	**224416**	227769.6	230704	1.49
tai12b	39464925	**39464925**	**39464925**	**39464925**	**0.00**
tai15a	388214	**388214**	388643.06	389718	0.11
tai15b	51765268	**51765268**	**51765268**	**51765268**	**0.00**
tai20a	703482	705622	710536.9	717390	1.00
tai20b	122455319	**122455319**	122566157.8	123009513	0.09
tai25a	1167256	1171944	1181543,733	1191072	1.22
tai25b	344355646	**344355646**	344383574.2	344565108	0.00
tai30a	1818146	1825262	1838228,4	1849970	1.10
tai30b	637117113	**637117113**	641132438.2	651247567	0.63

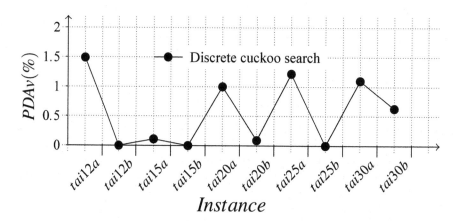

Fig. 5.13 PDav(%) of 10 QAP benchmark instances solved by DCS

pendent validation of DCS. These results confirm, through Table 5.6 and Fig. 5.13, the performance of this first version of the direct adaptation of DCS to QAP. DCS is able to converge on promising areas containing the best solutions while providing minimal effort.

5.5   Conclusion

In this chapter, DCS is adapted and applied to solve a well-selected set of combinatorial optimization problems. Its performance is tested on a set of benchmark instances of TSP, JSSP, and QAP, in order to verify how DCS manages the balance between intensification and diversification through Lévy flights and the new population structure that contains a variety of cuckoos adopting a multitude of research methods. However, the independence of the best solution of our cuckoo category can present a good strategy to generate new good solutions without considering the optimal solution of the population.

In harmony with this first validation of DCS, the central idea of this application is to take a step toward the generalization of DCS to solve more models of combinatorial optimization problems and to foresee potential improvements adapted to the different variants of these problems. This is encouraged through the results of the tests obtained and reported.

Following these applications, we see that DCS is leading an exploration of the global search space and exploitation of promising regions on different topologies of the search space. DCS component functionality independence allows integration of advanced local search methods for targeted performance based on the specificity of the problem being addressed.

References

Applegate D, Cook W (1991) A computational study of the job-shop scheduling problem. ORSA J Comput 3(2):149–156

Buaklee W, Hongesombut K (2013) Optimal dg allocation in a smart distribution grid using cuckoo search algorithm. In: 2013 10th international conference on electrical engineering/electronics, computer, telecommunications and information technology. IEEE, pp 1–6

Burkard RE, Karisch SE, Rendl F (1997) Qaplib-a quadratic assignment problem library. J Glob Optim 10(4):391–403

Croes A (1958) A method for solving traveling salesman problems. Oper Res: 791–812

Davendra D, Zelinka I (2012) Optimization of quadratic assignment problem using self organising migrating algorithm. Comput Inform 28(2):169–180

Fister I Jr, Yang X-S, Fister D, Fister I (2014) Cuckoo search: a brief literature review. In: Cuckoo search and firefly algorithm. Springer, pp 49–62

Guo P, Cheng W, Wang Y (2015) Parallel machine scheduling with step-deteriorating jobs and setup times by a hybrid discrete cuckoo search algorithm. Eng Optim 47(11):1564–1585

Koopmans TC, Beckmann M (1957) Assignment problems and the location of economic activities. Econ: J Econ Soc:53–76

Mahmoudi S, Lotfi S (2015) Modified cuckoo optimization algorithm (mcoa) to solve graph coloring problem. Appl Soft Comput 33:48–64

Manne AS (1960) On the job-shop scheduling problem. Oper Res 8(2):219–223

Mishra C, Singh SP, Rokadia J (2015) Optimal power flow in the presence of wind power using modified cuckoo search. IET Gener Transm Distrib 9(7):615–626

Ouaarab A, Ahiod B, Yang X-S (2014) Discrete cuckoo search algorithm for the travelling salesman problem. Neural Comput Appl 24(7–8):1659–1669

Reinelt G (1991) Tspliba traveling salesman problem library. ORSA J Comput 3(4):376–384

Sahni S, Gonzalez T (1976) P-complete approximation problems. J ACM (JACM) 23(3):555–565

Shehab M, Khader AT, Al-Betar MA (2017) A survey on applications and variants of the cuckoo search algorithm. Appl Soft Comput 61:1041–1059

Taillard E (1991) Robust taboo search for the quadratic assignment problem. Parallel Comput 17(4):443–455

Valian E, Tavakoli S, Mohanna S, Haghi A (2013) Improved cuckoo search for reliability optimization problems. Comput Ind Eng 64(1):459–468

Wang H, Wang W, Sun H, Cui Z, Rahnamayan S, Zeng S (2017) A new cuckoo search algorithm with hybrid strategies for flow shop scheduling problems. Soft Comput 21(15):4297–4307

Yang XS, Deb S (2009) Cuckoo search via lévy flights. In: World congress on nature and biologically inspired computing, 2009. NaBIC 2009. IEEE, pp 210–214

Yang X-S, Deb S (2014) Cuckoo search: recent advances and applications. Neural Comput Appl 24(1):169–174

Zheng HQ, Zhou Y, Luo Q (2013) A hybrid cuckoo search algorithm-grasp for vehicle routing problem. J Converg Inf Technol 8(3)

Chapter 6
Random-Key Cuckoo Search (RKCS) Applications

This chapter represents the random-key encoding scheme originally performed by Bean (1994) to solve, using genetic algorithms, combinatorial optimization problems for which the solutions are a sequence of integers. It is based on the process of interpreting random real numbers from the continuous space to encode a solution for a given combinatorial space. Based on the approach of Ouaarab et al. (2015a). It is combined random keys with cuckoo search to solve combinatorial problems such as traveling salesman and quadratic assignment problems (Ouaarab et al. 2015b). The idea is to try the transition from continuous to discrete space by avoiding the passage of traditional adaptation operators that can affect the performance of the algorithm and to rely firmly on a direct interpretation of operators used by metaheuristics in the continuous search space.

It is principally a projected version of CS on the combinatorial space. It lets CS searches and generates solutions in the continuous space and then makes a projection via random keys on the combinatorial space. After describing the main tools of Random-Key Cuckoo Search (RKCS), the remainder of the chapter consists of two CS applications by the random keys. The first is an application on the TSP, while the second concerns the resolution of QAP. Each of these applications describes in detail its adaptation procedure and confirms the approach with performance tests and a discussion of the obtained numerical results (Feng et al. 2014).

6.1 RKCS

Random-key encoding scheme (Bean 1994) is a widely used technique for transforming positions of the continuous space into combinatorial positions. It uses a vector of real numbers by associating each number with its weight. These weights will be used to generate a combination as a solution.

© Springer Nature Singapore Pte Ltd. 2020   71
A. Ouaarab, *Discrete Cuckoo Search for Combinatorial Optimization*,
Springer Tracts in Nature-Inspired Computing,
https://doi.org/10.1007/978-981-15-3836-0_6

Fig. 6.1 Random-key
encoding scheme

Random keys:	0.8	0.5	0.7	0.1	0.4	0.2

Decoded as:	6	4	5	1	3	2

The random real numbers have values in [0, 1] and constitute a vector like, for example, that represented in Fig. 6.1. On the other side, the combinatorial vector is composed of integers ordered according to the weights of real numbers in the first vector.

To solve combinatorial optimization problems, this approach offers a tool with a significant level of efficiency and ease of implementation, especially for most of meta-heuristics that are designed principally for continuous space and find difficulties to be adapted to the combinatorial space. This approach avoids many adaptation methods and let the metaheuristic move freely in its continuous space. In CS case, looking for new cuckoos by Lévy flights displacements is performed with real numbers. After finding a position, it is transformed by post-processing random-key encoding scheme to its corresponding combinatorial position.

6.1.1 Solution Representation

Random-key cuckoo search approach is mainly guided by two search strategies: (1) global search carried out at the level of the solution regions, controlled by the movements (according to the Lévy flights) of agents; and (2) local search that detects the best solutions in regions found by agents. The combination of both local and global searches results in improved performance.

Figure 6.2 presents the procedure of generating a combinatorial solution (a series of integers) using the random keys. First, the agents are randomly positioned according to their values in [0, 1]. Each agent has an integer (index) regardless of its increasing order among the other agents in the chained list (Fig. 6.2). All agents are ordered according to their weights (real numbers) and their indices form, together an initial solution. So, this essentially means that the integers/indices of the agents correspond to the discrete solution variables.

6.1.2 Displacement

RKCS approach extends the random-key mechanism by integrating Lévy flights to generate the random numbers (Ouaarab et al. 2015a). This makes it possible to directly balance the search for a solution in the local areas as well as at the global level.

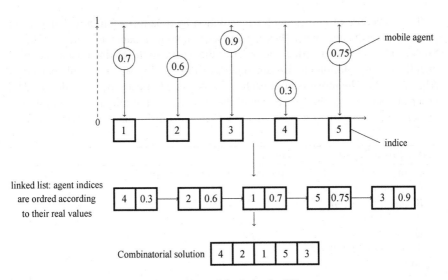

Fig. 6.2 Generation procedure of a combinatorial solution by RK

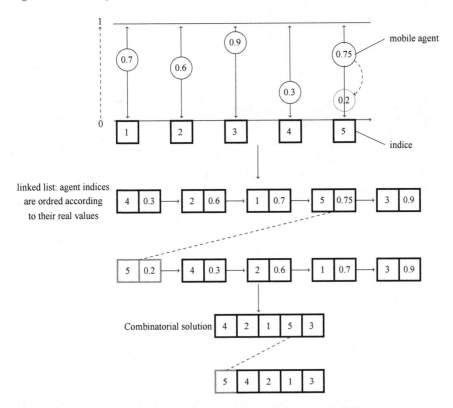

Fig. 6.3 Displacement procedure toward a new combinatorial solution by RK

The procedure of generating a new solution by a perturbation in real space can produce some changes when these agents start moving on bars with values in [0, 1]. Such displacements can affect the order of the agents in the chained list and consequently, a new solution can be generated as shown in Fig. 6.3 and implemented in Algorithm 24. The movement of each agent is guided directly by Lévy flights. The order of each city is changed by small disturbances or big jumps according to the values generated in their weight via Lévy flights as implemented in Algorithm 23.

Algorithm 23: City perturbation

Input: Lévy parameter: λ, predecessor and successor weights: $rnd1$ and $rnd2$
bar=random;
if $bar > 0.5$ **then**
| $\alpha = rnd2+$Lévy(λ)
else
| $\alpha = rnd1-$Lévy(λ)
end
   /* $\alpha \in [0, 1]$ */
if $\alpha > 1$ **then**
| $rnd = 2 - |\alpha|$
else
| $rnd = |\alpha|$
end

Algorithm 24: TSP Solution perturbation

Input: City order: i, Lévy parameter: λ
Get $rnd1$ the weight of the predecessor of i;
Get $rnd2$ the weight of the successor of i;
Perturb the weight of i out of $[rnd1, rnd2]$;
Place i in its new place in the list;

6.1.3 Lévy Flights

To approximate the step lengths guided by Lévy flights, RKCS has used Mantegna's algorithm step length s as calculated by

$$s = \frac{u}{|v|^{\frac{1}{\beta}}}, \qquad (6.1)$$

where u and v can be drawn from normal distributions $N(0, \sigma_u{}^2)$ and $N(0, \sigma_v - 2)$ (Yang 2010), where

$$\sigma_u = \left\{ \frac{\Gamma(1+\beta)\sin(\pi\beta/2)}{\Gamma[(1+\beta)/2]\beta 2^{(\beta-1)/2}} \right\}^{\frac{1}{\beta}}, \ \sigma_v = 1. \tag{6.2}$$

Γ is the gamma function that coincides with ordinary factorials $z!$ if z is integer according to the equation

$$\Gamma(z+1) = z!. \tag{6.3}$$

As contributed in the work of Lanczos (1964), $z!$ can be expressed as follows:

$$z! = (z+\gamma+1/2)^{z+1/2} \exp^{-(z+\gamma+1/2)} \sqrt{2\pi}[A_\gamma(z) + \varepsilon]. \tag{6.4}$$

For a $\gamma = 5$, $A_5(z)$ will equal to

$$A_5(z) = 1.000000000178 + \frac{76.180091729400}{z+1} - \frac{86.505320327112}{z+2} + \frac{24.014098222230}{z+3}$$
$$- \frac{1.231739516140}{z+4} + \frac{0.001208580030}{z+5} - \frac{0.000005363820}{z+6},$$
$$|\varepsilon| < 2.10^{10}. \tag{6.5}$$

Mantegna's approximation is implemented by Algorithm 25.

Algorithm 25: Lévy distribution

Input: λ
Input: s the step length
Let s be the step length;
Let G be a Gaussian variable;
$s = \frac{G \times \sigma(\lambda)}{|G|^{1/\lambda}}$;
Confine s in $[0, 1[$;

6.1.4 Main Functions

By adopting the same steps as those of improved CS (Ouaarab et al. 2014), the algorithm will begin by randomly generating the initial population. Its solutions are built according to the procedure illustrated in Fig. 6.2.

The second phase is launching a search, by a cuckoo around the best solution. The fitness of this cuckoo is compared to an individual of the population, chosen randomly. The best of the both will take a nest in the population. In this cuckoo search around the best solutions, CS uses Random Key (RK) to make moves inter-regions and then looks for the local optimum in the current region.

The third phase is a search by a portion p_c of smart cuckoos. These cuckoos begin by exploring new regions from their current solution. As shown in Fig. 6.3, they use Lévy flights to move in real space and as a result these displacements are interpreted to have a new combinatorial solution.

The last phase is for poor quality solutions that are abandoned to let place to other new solutions. RKCS starts with the search, for a new good solution, far from the best solution by a big jump. In this case, a big jump disrupts more agents via the Lévy flights. All these phases are presented and illustrated by a pseudocode with their main procedures in this section.

6.1.4.1   Get Cuckoo

Algorithm 26 perturbs a solution "Cuckoo" by random keys as described in Algorithm 27. In this perturbation, a random number of agents are perturbed to displace the "Cuckoo" to a new region. It moves in this found region toward a local optimum after performing a local search procedure. The best fitness between "Cuckoo" and "ind" is placed in the population.

Algorithm 26: Get cuckoo

Input: λ
Get the best solution of the population, Cuckoo=Sol;
PerturbList(Cuckoo,λ);
/* Random-key move   */
Cuckoo=localSearch(Cuckoo);
Get *ind*, a random individual in the population;
if *Fitness(Cuckoo)<Fitness(ind)* **then**
| replace *ind* by Cuckoo;
end

Algorithm 27: Get cuckoo

Input: λ
Result: Sol
Let Sol be a solution of *Dim* agent;
Get a random number of agents x;
Create a list of x agents;
for $i \leftarrow 0$ **to** *Dim* **do**
| Perturb agent i of Sol;
end

6.1.4.2 Smart Cuckoos

Apparently, smart cuckoo procedure (Algorithm 28) is similar to get cuckoo search procedure. However, the first one gives more diversity to the population. It avoids searching in the region of the best solution. The search is guided around the current solution. This procedure is applied on a portion p_c of smart cuckoos of the population.

Algorithm 28: Smart cuckoo

Input: λ,Sol
SCuckoo=Sol; Perturb SCuckoo by 27;
SCuckoo=localSearch(SCuckoo);
if *Fitness(SCuckoo)<Fitness(Sol)* **then**
  | Replace Sol by SCuckoo;
end

6.1.4.3 Worst Cuckoos

A portion p_a of worst cuckoos is replaced if the neighborhood is good enough to take their place in the population. The replacement of each worst cuckoo is controlled by the procedure of Algorithm 29.

Algorithm 29: Worst cuckoo

Input: λ,Sol
WCuckoo= the best; Perturb WCuckoo by 27;
WCuckoo=localSearch(WCuckoo);
if *Fitness(WCuckoo)< Fitness(Sol)* **then**
  | Replace Sol by WCuckoo;
end

6.2 Application on TSP

6.2.1 TSP Solution

RKCS approach builds a solution by a sorted list of cities as shown in Fig. 6.2. Each city is composed of a random real value $rnd \in [0, 1]$ and an $andex \in 0, \dots, n-1$ that represents the city number. rnd is a variable and controlled by Lévy flights. It is also used to localize where to insert the city in the list regarding its order among other city rnd values (Algorithm 30).

Algorithm 30: TSP Solution initialization

Result: Sol
for $i \leftarrow 0$ **to** Dim **do**
  $\mid$ rnd = random value;
  $\mid$ $index$ = i;
  $\mid$ Create a city c with rnd and $index$;
  $\mid$ Insert c according to the order of its rnd;
end

6.2.2 Displacement

Briefly, RKCS begins with a search for promising new areas. It combines intensification and diversification through small steps and big jumps to distant regions. After selecting an region found by Lévy flights, the best solution in this region is detected, and further research is initiated to locate a new area, which is described in the steps of Algorithm 24.

If rnd represents the continuous side of RKCS where all continuous operators are applied on rnd, $index$ represents the combinatorial side of the algorithm and is controlled by discrete local search methods.

6.2.3 Local Search and Neighborhood

In each found region, a local search is performed to find the best solution. This local search uses 2-opt (Croes 1958) moves as described in Sect. 5.2.3.1. In the case of minimization, the movement is performed only if the new tour is shorter than the current one. Obviously, these steps are repeated until no improvement is possible or a predefined number of iterations is reached.

The steepest descent (Algorithm 32) is a simple local search method, which tries to find a direction of descent toward the best solution in the neighborhood and which is often stuck in a local minimum and generally is not able to find a good quality solution. So, we chose this simple method to find the local optimum of a given region found by RK displacements. This allows us to generate good solutions, without introducing an advanced local search method that needs a considerable calculation time.

As described in Algorithm 31, a new solution is returned from the current one by applying one 2-opt move which is considered as a small step. After that, this method

is called in the local search function as shown in Algorithm 23. It calls successively
the first improving small step to reach the local optimum.

Algorithm 31: Neighbor step

Input: Sol, two city orders p_1 and p_2
Result: newSol
newSol=Sol;
newSol=$2 - opt(newSol, p_1, p_2)$;

Algorithm 32: RKCS local seach

Input: Sol
Result: neighbor
neighbor=Sol;
for $i, j \leftarrow 0$ **to** Dim **do**
  Sol=Neighbor step(neighbor,i, j);
  if *fitness(neighbor)>fitness(Sol)* **then**
    neighbor=Sol;
  end
end

6.2.4 Experimental Results

RKCS implemented and tested on some TSP benchmark instances of the TSPLIB
(Reinelt 1995). For each instance, 30 independent executions are performed. The
values of the parameters used in the experimental tests on RKCS, selected properly,
are shown in Table 6.1.

Table 6.2 shows the test results by RKCS execution for the resolution of a set
of 11 benchmark instances of the symmetric TSP. The first column is the name
of the instances with their optimums in parentheses. The "best" column shows the

Table 6.1 RKCS parameters

Parameter	Value	Signification
n	30	Population size
p_c	0.6	Portion of smart cuckoos
p_a	0.2	Portion of worst solutions
$MaxGen$	500	Maximum number of iterations
α	0.01	Step size
λ	1	Index

Table 6.2 Experimental results of the cuckoo search algorithm by random key (RKCS) for the TSP

Instance (opt)	Best	Average	Worst	PDav (%)
eil51(426)	**426**	426.9	430	0.21
berlin52(7542)	**7542**	7542	7542	0.0
st70(675)	**675**	677.3	684	0.34
pr76(108159)	**108159**	108202	109085	0.03
eil76(538)	**538**	539.1	541	0.20
kroA100(21282)	**21282**	21289.65	21343	0.03
eil101(629)	**629**	631.1	636	0.33
bier127(118282)	**118282**	118798.1	120773	0.43
pr136(96772)	97046	97708.9	98936	0.96
pr144(58537)	**58537**	58554.45	58607	0.02
ch130(6110)	6126	6163.3	6210	0.87

value/cost of the best solution found by RKCS, the "average" column gives the average solution found by RKCS, and the "worst" column shows the length of the worst solution of the RKCS independent executions.

These results confirm that the proposed approach is able to find good or optimal quality solutions for the proposed benchmark instances. Therefore, we can confirm that the random-key scheme can be very useful for moving from continuous space to the space of combinations. It also allows the move methods of the continuous space to behave freely and to project the produced changes by these methods into the combinatorial space. This approach facilitates controlling the balance between intensification and diversification through Lévy flights, which takes small incremen-

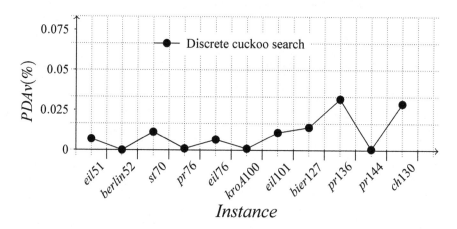

Fig. 6.4 PDav(%) of 11 TSP instances solved by RKCS

tal steps in a limited area followed by a large exploring jump to a distant region. Using real numbers, Lévy flights can easily act over the combinatorial distance and can directly generate small or large steps in the combinatorial space, with the help of random keys that project Lévy flights perturbations in the space of the TSP solutions (Fig. 6.4).

6.3 Application on QAP

6.3.1 QAP Solution

A solution of the Quadratic Assignment Problem (QAP) is a vector of N integers. Each integer is the index of a facility, and its order in the vector is the index of the corresponding location.

Algorithm 33: QAP Solution initialization

Result: Sol

Let Sol be an empty list of facilities;

for $i \leftarrow 0$ **to** Dim **do**

$\quad$ rnd = random value;

$\quad$ $index$ = i;

$\quad$ Create a facility c with rnd and $index$;

$\quad$ Insert c according to the order of its rnd;

end

6.3.2 Displacement

Considering Fig. 6.1, we can say that the resulting vector is a solution of the QAP. So, to move from the current solution to a new solution, we just have to perturb the first vector containing the real numbers. This perturbation is carried out via Lévy flights, which promotes the balance of finding solutions in local regions as well as in global regions.

6.3.3 Local Search

To detect the best solution in a region, a simple local search (Avriel 2012) is adopted. In this local search method, we used swaps as shown in Fig. 6.5 (In this example, we have fixed the choice on facilities of location 2 and 6, which are 1 and 9.) to move from a ϕ position to its neighbor π by applying a swap between r and s facilities:

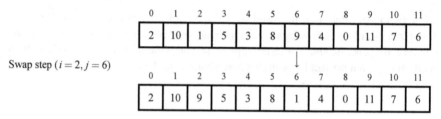

Swap step ($i = 2, j = 6$)

Fig. 6.5 Displacement swap operator

$$\pi(k) = \phi(k), \quad \forall k \neq r, s \tag{6.6}$$

$$\pi(r) = \phi(s), \ \pi(s) = \phi(r) \tag{6.7}$$

Before each move with a simple swap, the cost estimation of $\delta(\phi, r, s)$ uses the following formula:

$$\delta(\phi, r, s) \quad = \sum_{i=1}^{N} \sum_{j=1}^{N} (a_{ij} b_{\phi(i)\phi(j)} - a_{ij} b_{\pi(i)\pi(j)}) \tag{6.8}$$

$$= 2. \sum_{k \neq r, s} (a_{sk} - a_{rk})(b_{\phi(s)\phi(k)} - b_{\phi(r)\phi(k)}). \tag{6.9}$$

In the case of minimization, the swap is applied only if the cost of the new solution is less than that of the current solution. Obviously, this procedure is repeated as long as improvements are possible or before meeting the stopping criterion.

For the same reasons, QAP makes the same choice of simple local search method as for TSP to show the performance of CS combined with RK during the resolution of QAP instances. This allows us to generate good quality solutions, without the introduction of an advanced local search method. This local search method as implemented in Algorithm 34 moves in the neighborhood of the current solution *Sol* by using swap operator presented in Algorithm 35.

Algorithm 34: RKCS local seach

Input: Sol
Result: neighbor
neighbor=Sol;
for $i, j \leftarrow 0$ **to** *Dim* **do**
  Sol=swapStep(neighbor,*i*, *j*);
  if *fitness(neighbor)>fitness(Sol)* **then**
    neighbor=Sol;
  end
end

Algorithm 35: Neighbor step

Input: Sol, two facilities c_1 and c_2
Result: newSol
newSol=Sol;
newSol=swap($newSol, c_1, c_2$;

6.3.4 Experimental Results

We will show in this part the test results of RKCS for the resolution of instances (Skorin-Kapov 1990; Taillard 1991, 1995; Christofides and Benavent 1989; Burkard and Offermann 1977; Elshafei 1977; Krarup and Pruzan 1978; Wilhelm and Ward 1987; Nugent et al. 1968) QAP from the Quadratic Assignment Problem Library (QAPLIB) (Burkard and Offermann 1977). Twenty-eight instances are considered with sizes ranging from 12 up to 35 facilities. The numeric value in the instance name represents the number of facilities provided, e.g., the instance named sko90 has 90 facilities. We note that, for each instance, 10 independent executions are performed.

We implemented RKCS (Table 6.3) using the Java language on the MS Windows Seven 32-bit operating system. The experiments are conducted on a laptop with Intel [R] Core [TM] 2 Duo 2.00 GHz CPU, and 2 GB of RAM.

In the experimental result table, the first "Instance" column shows the name of the instance, while the "Bkv" column shows the value of the optimal solution from QAPLIB (Burkard et al. 1997). Table 6.4 which reports the results of the RKCS test statistics on some QAP instances contains nine columns. The "best" column shows the value of the best solution found. The "average" column gives the average value of the solutions of 10 independent executions, and the "worst" column shows the value of the worst solution found by RKCS. The "SD" column denotes the standard deviation, and the "PDav(%)" column shows the percentage deviation of the average solution value from the average solution value of 10 executions. The numbers in bold indicate that the solutions found have the same values as Best known value (Bkv).

Table 6.3 RKCS parameters

Parameter	Value	Signification
n	20	Population size
p_c	0.6	Portion of smart cuckoos
p_a	0.2	Portion of worst solutions
$MaxGen$	500	Maximum number of iterations
α	0.01	Step length
λ	1	Index
ρ	$Levy$	Mutation rate

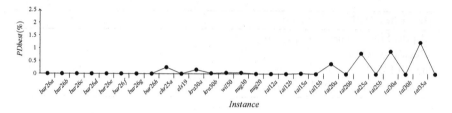

Fig. 6.6 PDbest(%) of 16 QAPLIB instances

The percentage deviation of the solution from the best known solution is given by Eq. 6.10.

$$PDsolution(\%) = \frac{solution\ value - best\ known\ solution\ value}{best\ known\ solution\ value} \times 100.$$

(6.10)

These results (Table 6.4 and Fig. 6.6) show that RKCS can be easily adapted to solve the QAP. However, we can confirm that random keys play an important role in the transition from continuous space to combinatorial space. This is in the favor of moving operators to behave freely and, therefore, facilitates the task of projecting changes made in continuous space. Through real numbers, Lévy flights are more compatible with distance notion, and they have more possibilities to clearly define a small or big step, and subsequently RK deals with the projection of these changes on the QAP solution space.

6.4 Conclusion

Random keys are designed to make the transition from continuous to combinatorial space more flexible and straightforward. The objective, therefore, is to facilitate the adaptation of metaheuristics while resolving combinatorial optimization problems. In the case of cuckoo search, exploratory displacement is one of the qualities of Lévy flights, and the role of RK is to offer a direct projection of these movements into the space of combinations. This approach allows Lévy flights to control fluently the balance between intensification and diversification.

Random-key cuckoo search has been applied for solving the traveling salesman problem and the quadratic assignment problem. The numerical results of the tests are encouraging. This second approach is carried out in the sense of facilitating the reuse of nature-inspired metaheuristics in a direct adaptation to solve COPs while keeping the basic structure of these metaheuristics and without considering the details of the problem dealt with.

Table 6.4 Experimental results of applying RKCS on a set of QAP benchmark instances

Instance	Bkv	Best	Average	Worst	SD	PDav(%)	PDbest(%)	Time(s)
bur26a	5426670	5426670	5426670.00	5426670	0.0	0.0	0.0	0.55
bur26b	3817852	3817852	3817852.00	3817852	0.0	0.0	0.0	0.76
bur26c	5426795	5426795	5426795.00	5426795	0.0	0.0	0.0	1.16
bur26d	3821225	3821225	3821225.00	3821225	0.0	0.0	0.0	1.77
bur26e	5386879	5386879	5386879.00	5386879	0.0	0.0	0.0	1.18
bur26f	3782044	3782044	3782044.00	3782044	0.0	0.0	0.0	0.60
bur26g	10117172	10117172	10117172.00	10117172	0.0	0.0	0.0	1.73
bur26h	7098658	7098658	7098658.00	7098658	0.0	0.0	0.0	0.75
chr25a	3796	3796	3805.75	3874	9.75	0.25	0.0	13.76
els19	17212548	17212548	17212548.00	17212548	0.0	0.0	0.0	0.22
kra30a	88900	88900	89048.75	90090	148.75	0.16	0.0	9.56
kra30b	91420	91420	91446.25	91490	26.25	0.02	0.0	30.05
wil50	48816	48816	48839.75	48886	23.75	0.04	0.0	220.50
nug30	6124	6124	6126.50	6128	2.5	0.04	0.0	39.90
nug20	2570	2570	2570.00	2570	0.0	0.0	0.0	0.67
tai12a	224416	224416	224416	224416	0.0	0.0	0.0	0.02
tai12b	39464925	39464925	39464925	39464925	0.0	0.0	0.0	0.03
tai15a	388214	388214	388224.8	388250	12.97	0.02	0.0	4.16
tai15b	51765268	51765268	51765268	51765268	0.0	0.0	0.0	0.06
tai20a	703482	703482	706342.0	708654	2565.75	0.40	0.0	19.94
tai20b	122455319	122455319	122455319	122455319	0.0	0.0	0.0	0.44
tai25a	1167256	1167256	1176732.8	1181894	7316.99	0.81	0.0	39.45
tai25b	344355646	344355646	344355646	344355646	0.0	0.0	0.0	1.13
tai30a	1818146	1827982	1834456.2	1843032	6474.19	0.89	0.54	74.01
tai30b	637117113	637117113	637117113	637117113	0.0	0.0	0.0	4.93
tai35a	2422002	2444344	2452223.4	2460532	7270.30	1.24	0.9	95.63
tai35b	283315445	283315445	283315445	283315445	0.0	0.0	0.0	10.52
tai40a	3139370	3164372	3178239.4	3192798	11917.91	1.23	0.79	140.86

References

Avriel M (2012) Nonlinear programming: analysis and methods. Courier Dover Publications

Bean J (1994) Genetic algorithms and random keys for sequencing and optimization. ORSA J Comput 6:154–154

Burkard RE, Karisch SE, Rendl F (1997) Qaplib-a quadratic assignment problem library. J Global Optim 10(4):391–403

Burkard RE, Offermann DMJ (1977) Entwurf von schreibmaschinentastaturen mittels quadratischer zuordnungsprobleme. Zeitschrift für Oper Res 21(4):B121–B132

Christofides N, Benavent E (1989) An exact algorithm for the quadratic assignment problem on a tree. Oper Res 37(5):760–768

Croes A (1958) A method for solving traveling salesman problems. Oper Res:791–812

Elshafei AN (1977) Hospital layout as a quadratic assignment problem. J Oper Res Soc 28(1):167–179

Feng Y, Jia K, He Y (2014) An improved hybrid encoding cuckoo search algorithm for 0–1 knapsack problems. In: Computational intelligence and neuroscience, 2014

Krarup J, Pruzan PM (1978) Computer-aided layout design. In: Mathematical programming in use. Springer, pp 75–94

Lanczos C (1964) A precision approximation of the gamma function. J Soc Ind Appl Math Ser B Numer Anal 1(1):86–96

Nugent CE, Vollmann TE, Ruml J (1968) An experimental comparison of techniques for the assignment of facilities to locations. Oper Res 16(1):150–173

Ouaarab A, Ahiod B, Yang X-S (2014) Improved and discrete cuckoo search for solving the travelling salesman problem. In: Yang X-S (eds) Cuckoo search and firefly algorithm. Studies in computational intelligence, vol 516. Springer International Publishing, pp 63–84

Ouaarab A, Ahiod B, Yang X-S (2015a) Random-key cuckoo search for the travelling salesman problem. Soft Comput 19(4):1099–1106

Ouaarab A, Ahiod B, Yang X-S, Abbad M (2015b) Random-key cuckoo search for the quadratic assignment problem (submitted). Nat Comput

Reinelt G (1995) Tsplib 1995. Universitat Heidelberg

Skorin-Kapov J (1990) Tabu search applied to the quadratic assignment problem. ORSA J Comput 2(1):33–45

Taillard E (1991) Robust taboo search for the quadratic assignment problem. Parallel Comput 17(4):443–455

Taillard ED (1995) Comparison of iterative searches for the quadratic assignment problem. Locat Sci 3(2):87–105

Wilhelm MR, Ward TL (1987) Solving quadratic assignment problems by simulated annealing. IIE Trans 19(1):107–119

Yang XS (2010) Nature-inspired metaheuristic algorithms. Luniver Press

Appendix A
Benchmarks

All's well that ends well.

In order to validate their performance, the developed combinatorial optimization algorithms are tested against a well-selected problem instances. In this purpose, DCS and RKCS have used, for test experiments a subset of TSP, JSSP, and QAP benchmark instances.

In this appendix, one example of each subset is provided. It is treated by DCS and RKCS similarly as all instances of the same subsets that have similar characteristics.

A.1 TSP

TSP instances are provided by TSPLIB, which contains a publicly available set of TSP problem data. The use of its instances is guided by documentation of Reinelt (1995). Each TSP file contains one problem instance. It is composed of two parts. The first part is for information related to the instance and the second is a list of cities coordinates. Information part contains the following indications:

- NAME: The name of the instance problem,
- COMMENT: The information such as number of cities and the author name,
- DIMENSION: The problem dimension (number of cities), and
- $EDGE\_WEIGHT\_TYPE$: How the edges weight is calculated.

The first column of the second part contains the numbers of cities, and the second and the third columns contain the city coordinates (x, y).

In DCS and RKCS experimental test instances have been selected from "EUC_2D" category. Edge weights of this category are euclidean distances in 2-D and calculated regarding the formula as follows:

© Springer Nature Singapore Pte Ltd. 2020
A. Ouaarab, *Discrete Cuckoo Search for Combinatorial Optimization*,
Springer Tracts in Nature-Inspired Computing,
https://doi.org/10.1007/978-981-15-3836-0

$$xd = x[i] - x[j]; \; yd = y[i] - y[j]; \; dij = nint(sqrt(xd * xd + yd * yd)); \quad \text{(A.1)}$$

For the rounding function "nint", "nint(x)" is replaced by "(int) (x + 0.5)".

```
NAME : eil51
COMMENT : 51-city problem (Christofides/Eilon)
TYPE : TSP
DIMENSION : 51
EDGE_WEIGHT_TYPE : EUC_2D
NODE_COORD_SECTION
1 37 52
2 49 49
3 52 64
4 20 26
5 40 30
6 21 47
7 17 63
8 31 62
9 52 33
10 51 21
11 42 41
12 31 32
13 5 25
14 12 42
15 36 16
16 52 41
17 27 23
18 17 33
19 13 13
20 57 58
21 62 42
22 42 57
23 16 57
24 8 52
25 7 38
26 27 68
27 30 48
28 43 67
29 58 48
30 58 27
31 37 69
32 38 46
33 46 10
34 61 33
35 62 63
36 63 69
37 32 22
38 45 35
39 59 15
40 5 6
41 10 17
42 21 10
```

```
43 5 64
44 30 15
45 39 10
46 32 39
47 25 32
48 25 55
49 48 28
50 56 37
51 30 40
EOF
```

A.2   JSSP

JSSP instances are provided from OR-Library Beasley (1990) (http://people.brunel.
ac.uk/~mastjjb/jeb/orlib/files/jobshop1.txt). JSP files of the used instances from this
link contain three parts (two lines and one matrix) described as follows:

- A description of the instance problem,
- A line containing the number of jobs and the number of machines, and
- One line for each job, listing the machine number and processing time for each
  operation of the job. Machines are numbered starting from 0.

```
Fisher and Thompson 6x6 instance, alternate name (mt06)
6   6
2   1   0   3   1   6   3   7   5   3   4   6
1   8   2   5   4  10   5  10   0  10   3   4
2   5   3   4   5   8   0   9   1   1   4   7
1   5   0   5   2   5   3   3   4   8   5   9
2   9   1   3   4   5   5   4   0   3   3   1
1   3   3   3   5   9   0  10   4   4   2   1
```

A.3   QAP

QAP instances are taken from the Quadratic Assignment Problem Library Burkard
et al. (1997). QAP instances such as "tail12a" are presented as follows:

```
12

 0 27 85  2  1 15 11 35 11 20 21 61
27  0 80 58 21 76 72 44 85 94 90 51
85 80  0  3 48 29 90 66 41 15 83 96
 2 58  3  0 74 45 65 40 54 83 14 71
 1 21 48 74  0 77 36 53 37 26 87 76
```

```
15 76 29 45 77  0 91 13 29 11 77 32
11 72 90 65 36 91  0 87 67 94 79  2
35 44 66 40 53 13 87  0 10 99 56 70
11 85 41 54 37 29 67 10  0 99 60  4
20 94 15 83 26 11 94 99 99  0 56  2
21 90 83 14 87 77 79 56 60 56  0 60
61 51 96 71 76 32  2 70  4  2 60  0

 0 21 95 82 56 41  6 25 10  4 63  6
21  0 44 40 75 79  0 89 35  9  1 85
95 44  0 84 12  0 26 91 11 35 82 26
82 40 84  0 69 56 86 45 91 59 18 76
56 75 12 69  0 39 18 57 36 61 36 21
41 79  0 56 39  0 71 11 29 82 82  6
 6  0 26 86 18 71  0 71  8 77 74 30
25 89 91 45 57 11 71  0 89 76 76 40
10 35 11 91 36 29  8 89  0 93 56  1
 4  9 35 59 61 82 77 76 93  0 50  4
63  1 82 18 36 82 74 76 56 50  0 36
 6 85 26 76 21  6 30 40  1  4 36  0
```

The first line is the problem dimension, followed by flow and distance matrices (the order is not always the same).

References

Beasley JE (1990) Or-library: distributing test problems by electronic mail. J Oper Res Soc 41(11):1069–1072

Burkard RE, Karisch SE, Rendl F (1997) Qaplib-a quadratic assignment problem library. J Glob Optim 10(4):391–403

Reinelt G (1995) Tsplib, 1995. Universitat Heidelberg

Appendix B
Computer Codes

The computer code presented in this appendix is a set of implementations of DCS on TSP, JSSP, and QAP; and RKCS on TSP and QAP. They intend to explain how DCS and RKCS perform their main techniques such as I/D related to the notion of step length and local search, cuckoo categories, and Lévy flights. Therefore, this implementation is simplified and represents the basic version, so it can be widely improved. The aim is to show how it is easy to apply DCS and RKCS by following the proposed model of adaptation that is based on nest, egg, objective function, and search space elements.

In this chapter, the first section contains a computer code implementation of DCS algorithm on TSP, JSSP, and QAP instances. Each implementation is organized into three classes, and each class represents an independent mechanism (function or procedure) of the program. The first (TSP, JSSP, or QAP) is for reading and using the instance file. The second (DCSTSP, DCSJSSP, or DCSQAP) is for functions related to the instance adaptations. The third DCS implements the functions of DCS independently to the problem instances.

The second section is composed of two implementations of RKCS on TSP and QAP. It proposes an alternative model of classes for adapting CS to the problem instances. The model is based on three hierarchic classes: the solution component (CITY or FACILITY), the solution individual (CityList or FacilityList), and the population (Population). RKCSTSP or RKCSQAP contain the main function that runs the program with respect to CS procedure.

© Springer Nature Singapore Pte Ltd. 2020
A. Ouaarab, *Discrete Cuckoo Search for Combinatorial Optimization*,
Springer Tracts in Nature-Inspired Computing,
https://doi.org/10.1007/978-981-15-3836-0

B.1   DCS for TSP

B.1.1   TSP

This class reads the TSP file, mainly the cities' coordinates and generates the distance matrix "matrix" that contains distances between each couple of cites. It also provides the fitness function "fitness" that returns the length of a solution tour.

```java
package dcstsp;
import java.io.File;
import java.io.FileNotFoundException;
import java.util.ArrayList;
import java.util.Scanner;
import java.util.StringTokenizer;

public final class Tsp {
  private String name,
          comment="",
          type,
          edge_weight_type;
  private String dimension;
   //Contain coordinates of the cities
  public static ArrayList<ArrayList<Double>> node_coord_section=new
      ArrayList<>();
   //Contain distances between each pair of cities
  public static ArrayList<ArrayList<Integer>> matrix=new ArrayList<>();

  public Tsp(File f){
    info(f);
    coordinate(f);
    matrix=DistanceMatrix();
  }

  Tsp() { }

  // Reading city coordinates from 0 to n-1
  public void coordinate(File f){
    Scanner input1=null;
    int dex=0;
    try {
       input1=new Scanner(f);
    } catch (FileNotFoundException e) {
      System.out.println(e.toString());
    }
    while(input1.hasNextLine()){
      StringTokenizer str1=new StringTokenizer(input1.nextLine()," ");
      String str2=" ";
      if(str1.hasMoreTokens()) str2=str1.nextToken();
      if(str2.equalsIgnoreCase("EOF")||str2.equalsIgnoreCase("-1"))
        dex=0;
      if(dex==1){
        ArrayList<Double> s=new ArrayList<>();
        StringTokenizer no1=new StringTokenizer(str1.nextToken()," ");
        s.add(Double.valueOf(no1.nextToken()));
        StringTokenizer no2=new StringTokenizer(str1.nextToken()," ");
        s.add(Double.valueOf(no2.nextToken()));
```

```
        node_coord_section.add(s);
      }
    if(str2.equalsIgnoreCase("NODE_COORD_SECTION"))
        dex=1;
  }
}


//Reading instance information
public void info(File f) {
  Scanner input = null;
  try {
    input = new Scanner(f);
    }
  catch (FileNotFoundException e) {
        System.out.println(e.toString());
    }
  String info;
  while(input.hasNextLine()){
    StringTokenizer str=new StringTokenizer(input.nextLine(),":");
    while(str.hasMoreTokens()){
      info=str.nextToken();
      if(info.contains("NAME"))
        name=str.nextToken();
      if(info.contains("COMMENT"))
        comment+=str.nextToken()+"\n";
      if(info.equalsIgnoreCase("TYPE "))
        type=str.nextToken();
      if(info.contains("DIMENSION"))
        dimension=str.nextToken();
      if(info.contains("EDGE_WEIGHT_TYPE"))
        edge_weight_type=str.nextToken();
    }
  }
}

//Distance between two cities v1 and v2
public double distance(ArrayList<Double> v1,ArrayList<Double> v2){
    return Math.sqrt(Math.pow((v1.get(0)-v2.get(0)),2)
            +Math.pow((v1.get(1)-v2.get(1)),2));
}

//Distance matrix
public ArrayList<ArrayList<Integer>> DistanceMatrix(){
  ArrayList<ArrayList<Integer>> matrix=new ArrayList<>();
  for(int i=0;i<node_coord_section.size();i++){
    ArrayList<Integer> line=new ArrayList<>();
    for(int j=0;j<node_coord_section.size();j++){
      line.add((int)(distance(node_coord_section.get(i),
          node_coord_section.get(j))+0.5));
    }
    matrix.add(line);
  }
  return matrix;
}

//Tour fitness
public static int fitness(ArrayList<Integer> tour){
  int length=0;
```

```
    for(int i=0;i<dim();i++){
      longth+=matrix.get(tour.get(i)).get(tour.get(i+1));
    }
    return longth;
  }

  //The instance dimension
  public static int dim(){return node_coord_section.size(); }
}
```

B.1.2   DCSTSP

DCSTSP class is the link between TSP and DCS classes. It adapts DCS functions to be implemented on TSP instances.

```
package dcstsp;
import java.io.File;
import java.io.IOException;
import java.util.ArrayList;

public class DCSTSP{
   static int dim;
   static Tsp tsp;

   public static void main(String[] args) throws IOException{
     File f=new File("berlin52.tsp");
      tsp=new Tsp(f); dim=Tsp.dim();
      DCS dcs=new DCS(0.2,0.6,2,20,500);
     dcs.run();
     System.out.println(DCS.popCost.get(0));
   }

   //Local search around a solution ``sol''
   public static ArrayList<Integer> localSearch(ArrayList<Integer>
       sol) {
     int r=0;
     ArrayList<Integer> sol1=cloner(sol);
     for(int i=0;i<dim-2;i++){
        if(i==0) r=0;
        for(int j=i+2;j<dim-1+r;j++){
           if(gain(sol1,i,j)<0){
              sol1=smallStep(sol1, i, j);}
        }
        r=1;
        }
     return sol1;
   }

   //The gain of a 2-opt move between cities i and j in a solution sol
   private static int gain(ArrayList<Integer> sol, int i, int j) {
     int x=tsp.matrix.get(sol.get(i)).get(sol.get(i+1));
     int y=tsp.matrix.get(sol.get(j)).get(sol.get((j+1)%(dim)));
     int u=tsp.matrix.get(sol.get(i)).get(sol.get(j));
```

```
    int v=tsp.matrix.get(sol.get(i+1)).get(sol.get((j+1)%(dim)));
    return (u+v)-(x+y);
}

//Small step from a solution sol.
//It is performed by 2-opt move between a pair of random cities
public static ArrayList<Integer> smallStep(ArrayList<Integer> sol)
      {
    int p1=(int)(Math.random()*(dim-2));
    int r=p1==0?0:1;
    int p2=p1+2+(int)(Math.random()*(dim-p1-3+r));
    return smallStep(sol, p1 , p2);
}

//Small step (2-opt) between cities p1 and p2 in solution sol
public static ArrayList<Integer> smallStep(ArrayList<Integer> sol,
      int p1, int p2) {
    ArrayList<Integer> sol1=cloner(sol);
    for(int i=p1+1,j=p2;i<j;i++,j--){
      sol1.set(i, sol.get(j));
      sol1.set(j, sol.get(i));
  }
  return sol1;
}

//Big step from a solution sol
//It is performed by double-bridge move
public static ArrayList<Integer> bigStep(ArrayList<Integer> sol) {
    int p1,p2,p3;
    ArrayList<Integer> sol1=new ArrayList<>();
    p1=2 + (int)(Math.random()*(double)(dim-6));
    p2=p1 + 2 + (int)((Math.random()*(double)(dim-p1-4)));
    p3=p2 + 2 + (int)(Math.random()*(double)(dim-p2-2));
    for(int i=0;i<p1;i++)
       sol1.add(sol.get(i));
    for(int i=p3;i<dim;i++)
       sol1.add(sol.get(i));
    for(int i=p2;i<p3;i++)
       sol1.add(sol.get(i));
    for(int i=p1;i<p2;i++)
       sol1.add(sol.get(i));
    sol1.add(sol.get(dim));
    return sol1;
}

//Clone a solution
public static ArrayList<Integer> cloner(ArrayList<Integer> tour) {
    ArrayList<Integer> city=new ArrayList<>();
    for(int i=0;i<tour.size();i++){
      city.add(tour.get(i));}
    return city;
  }
}
```

B.1.3 DCS

DCS class is an interpretation of DCS algorithm implemented to be independent of the problem instances. It contains DCS main mechanisms.

```
package dcstsp;
import java.util.ArrayList;

public class DCS {
    public static ArrayList<ArrayList<Integer>> population=new
        ArrayList<>();
//Population solution costs (fitnesses)
  public static ArrayList<Integer> popCost=new ArrayList<>();
//Population size
  public static int n;
//Max generations
  public int maxgen;
  public double pa;
  public double pc;
  public double mu;

  public int dim;

  DCS(double a,double c, double m, int np, int max){
    pa=a; pc=c; mu=m; n=np; maxgen=max;
    dim=Tsp.dim();
  }

  public void run(){
    //Initialize the population
    population();
    //Sort the population: the best is in the first position
    sortPopulation();
    //Launch the algorithm
    while(--maxgen>0){
      cs();
      getCukoo();
      worst();
    }
  //Print the result
  System.out.println("result: "+popCost.get(0));
  }
  /////////////////////////******************************///////////////////////

  //Population method generates a random initial population with its
      population costs
  private void population() {
    for(int i=0;i<n;i++){
      population.add(solution());
      popCost.add(Tsp.fitness(population.get(i)));
    }
  }
```

```
//popCost contains the fitness value of each individial in the
     same order
private static void popCost() {
   popCost.clear();
   for(int i=0;i<n;i++){
     popCost.add(Tsp.fitness(population.get(i)));
   }
}

//Generate a random solution
private ArrayList<Integer> solution() {
  ArrayList<Integer> solution=new ArrayList<>();
  int temp,rnd;
  for(int i=0;i<dim;i++)
    solution.add(i);
  for(int j=0;j<dim;j=j+4){
    temp=solution.get(j);
    rnd=(int)(Math.random()*dim);
    solution.set(j, solution.get(rnd));
    solution.set(rnd, temp);
  }
  solution.add(solution.get(0));
  return solution;
}

//Sort the population respecting to the individual fitness value
private static void sortPopulation() {
   ArrayList<ArrayList<Integer>> newPop=new ArrayList<>();
   int i=0,j=0;
   newPop.add(j, DCSTSP.cloner(population.get(i++)));
   do{
     boolean inc=false;
     while(!inc && j<newPop.size()){
       if(popCost.get(i)<Tsp.fitness(newPop.get(j))){
         newPop.add(j, DCSTSP.cloner(population.get(i)));
         inc=true;
       }
       else j++;
     }
     if(!inc) newPop.add(j, DCSTSP.cloner(population.get(i)));
       j=0;
     i++;
   }while(i<n);
   population=newPop;
   popCost();
}

//Search for new nests with the smart portion of cuckoos
private void cs() {
  ArrayList<Integer> sol;
  double levy;
  for(int i=1;i<n*pc;i++){
```

```java
          levy=levy(mu);
        sol=DCSTSP.cloner(population.get(i));
        if(levy<0.5){
           sol=DCSTSP.localSearch(DCSTSP.smallStep(sol));}
        else
           sol=DCSTSP.localSearch(DCSTSP.bigStep(sol));
        if(popCost.get(i)>Tsp.fitness(sol))
             {population.set(i, sol);
              popCost.set(i, Tsp.fitness(sol));}
     }
      sortPopulation();
   }


   //Get a cuckoo and choose randomly an individual.
   //The better among the cuckoo and the individual takes a nest
       in the population
   private void getCukoo() {
     ArrayList<Integer> sol=DCSTSP.cloner(population.get(0));
     double levy=levy(mu);
     int k=(int) (Math.random()*(n-(n*pa)-(n*pc))+n*pa);
     if(levy<0.5){
        sol=DCSTSP.localSearch(DCSTSP.smallStep(sol));}
     else
        sol=DCSTSP.localSearch(DCSTSP.bigStep(sol));
     sol=DCSTSP.localSearch(sol);
     if(popCost.get(k)>Tsp.fitness(sol))
        {population.set(k, sol);}
     popCost();
     sortPopulation();
   }


   //Replace a portion p_a of worst individuals by new better
       cuckoos
   private void worst(){
      ArrayList<Integer> sol;
      for(int i=(int)(n-(n*pa));i<n;i++){
        sol=DCSTSP.cloner(population.get(0));
        sol=DCSTSP.localSearch(DCSTSP.bigStep(sol));
        sol=DCSTSP.localSearch(sol);
        if(popCost.get(i)>Tsp.fitness(sol))
          {population.set(i,sol);
             popCost.set(i, Tsp.fitness(sol));}
      }
      sortPopulation();
   }


   //A simple approximation of Levy flights distribution
   private static double levy(double mu){
     double t=Math.random()*9+1;//
       t=Math.pow((1/t),(mu));
     return t;
   }
}
```

B.2   DCS for JSSP

B.2.1   JSSP

Like TSP class, JSSP class reads the JSP file and generates the number of jobs
and machines (N and M), two matrices of operations and their processing times
(Machines and Durations). Its fitness function returns the makespan of the solution.

```
package dcsjssp;
import java.io.File;
import java.util.ArrayList;
import java.util.Scanner;

public class JSSP {
  static int N;//Number of jobs
  static int M;//Number of machines
  static int[][] Machines; //Machine operations
  static int[][] Durations; //Processing time

  //Read JSSP instance information
  public JSSP(File dataSource) {
    Scanner input = null;
     Scanner line = null;
      try {
       input = new Scanner(dataSource);
         int i=0, j=0, tache, operation;
         while (i < 2) {
         if(i==1) {
            line = new Scanner(input.nextLine());
            N = line.nextInt();
            M = line.nextInt();
            Machines = new int[N][M];
            Durations = new int[N][M];
         }
         else input.nextLine();
         i++;
         }
         i=0;
         while (input.hasNextLine()) {
            line = new Scanner(input.nextLine());
            tache = i; //
            for(j=0; j<M; j++) {
             operation = line.nextInt();
             Machines[tache][j] = operation;
             Durations[tache][j] = line.nextInt();
            }
            i++;
         }
     } catch (Exception e) { System.out.println(e.getMessage()); }
  }
```

```
//Fitness (Makespan) of a solution (schedule) sol
static int fitness(ArrayList<Integer> sol){
 ArrayList<Integer> machines=new ArrayList<>();
 ArrayList<Integer> jobs=new ArrayList<>();
 ArrayList<Integer> jind=new ArrayList<>();
 int max=0;
 for(int i=0; i<N; i++){
   jobs.add(0);
   jind.add(0);
 }
 for(int i=0; i<M; i++){
   machines.add(0);
 }
 int j,m,mach,d;
 for(int i=0; i<N*M; i++){
   //k is the job number starting from 0
   j=sol.get(i);
   //l is the operation number in the starting from 0
   m=jind.get(j);
   //mach is the current machine number
   mach=Machines[j][m];
   //Processing time in a given machine m of a job j
   d=Durations[j][m];
   machines.set(mach,Math.max(machines.get(mach),jobs.get(j))+d);
   jind.set(j, m+1);
   jobs.set(j, machines.get(mach));
 }
 for(int i=0; i<machines.size(); i++){
    if(max<machines.get(i)){
       max=machines.get(i);
    }
 }
 return max;
}
}
```

B.2.2 DCSJSSP

DCSJSSP class contains adaptation functions that allow DCS algorithm to be implemented on a JSSP instance. The adaptation is based on interpreting local search method with its different step moves in the JSSP solution space.

```
package dcsjssp;
import java.io.File;
import java.util.ArrayList;

public class DCSJSSP {
  //A variable that contains the instance information
    static JSSP jssp;
```

```java
//Instance dimension
static int dim;

public static void main(String[] args) {
  jssp= new JSSP(new File("ft06.jsp"));
   dim=jssp.N*jssp.M;
   DCS dcs=new DCS(0.2, 0.8, 2, 40, 30);
   //Run DCS
  dcs.run();
}


//Local search around a solution ''sol''
static ArrayList<Integer> localSearch(ArrayList<Integer> sol) {
  ArrayList<Integer> sol1=cloner(sol);
  ArrayList<Integer> sol2;
  for(int i=0;i<(int)(dim/2);i++){
    int k=(int)(Math.random()*dim);
    for(int j=0;j<(int)(dim/3);j++){
      int l=(int)(Math.random()*dim);
        sol2=perturbSwap(k,l,sol1);
        if(JSSP.fitness(sol1)>JSSP.fitness(sol2)){
          sol1=cloner(sol2);
          i=0;j=dim;}
    }
  }
  return sol1;
}
//Inverse the order in sol between position pos1 and pos2
public static ArrayList<Integer> invers(int pos1, int pos2,
    ArrayList<Integer> sol) {
  int temp=pos1;
  if(pos2<pos1)
    {pos1=pos2; pos2=temp;}
  for(int i=pos1, j=pos2; i<j; i++, j--){
    temp=sol.get(i);
    sol.set(i, sol.get(j));
    sol.set(j, temp);
  }
  return sol;
}

//Random inverse between two positions in sol
public static ArrayList<Integer> perturbInvers(ArrayList<Integer>
    sol) {
  int rnd1=(int)(Math.random()*(dim));
  int rnd2=(int)(Math.random()*(dim));
  return invers(rnd1 ,rnd2, sol);
}

//Random swap perturbation in sol
public static ArrayList<Integer> perturbSwap(ArrayList<Integer>
    sol){
  int rnd1=(int)(Math.random()*(dim));
  int rnd2=(int)(Math.random()*(dim));
```

```
    return perturbSwap(rnd1, rnd2, sol);
}

  //Random insertion perturbation in sol
  public static ArrayList<Integer> perturbInsert(ArrayList<Integer>
      sol){
    int rnd1=(int)(Math.random()*(dim));
    int rnd2=(int)(Math.random()*(dim));
    int tmp=rnd1;
    if(rnd2<rnd1){
      rnd1=rnd2; rnd2=tmp;}
    tmp=sol.get(rnd1);
    sol.remove(rnd1);
    sol.add(rnd2, tmp);
    return sol;
}

  //Swap move in sol between p1 and p2
  static ArrayList<Integer> perturbSwap(int p1, int p2,
      ArrayList<Integer> sol) {
    int temp=p1;
    if(p1>p2) {p1=p2; p2=temp;}
    temp=sol.get(p2);
    ArrayList<Integer> sol1=cloner(sol);
    sol1.remove(p2);
    sol1.add(p1, temp);
    return sol1;
}

//Clone a solution schedule
public static ArrayList<Integer> cloner(ArrayList<Integer>
    schedule) {
  ArrayList<Integer> l=new ArrayList<>();
  for(int i=0;i<schedule.size();i++){
    l.add(schedule.get(i)); }
  return l;
}


}
```

B.2.3   DCS

This class has the same structure as DCS class for TSP. The difference is inside its
functions where they call fitness and DCSJSSP functions.

```java
package dcsjssp;
import java.util.ArrayList;

public class DCS {
    public static ArrayList<ArrayList<Integer>> population=new
        ArrayList<>();
//Population solution costs (fitnesses)
    public static ArrayList<Integer> popCost=new ArrayList<>();
//Population size
    public static int n;
//Max generations
    public int maxgen;
    public double pa;
    public double pc;
    public double mu;

    public int job,machine;

    DCS(double a, double c, double m, int np, int max){
      pa=a; pc=c; mu=m; n=np; maxgen=max;
      job=JSSP.N; machine=JSSP.M;
    }

    public void run(){
      //Initialize the population
      population();
      //Sort the population: the best is in the first position
      sortPopulation();
      //Launch the algorithm
      while(--maxgen>0){
        cs();
        getCukoo();
        worst();
      }
    //Print the result
    System.out.println("result "+popCost.get(0));
    System.out.println("solution"+population.get(0));
    }

    /////////////////////////********************//////////////////////////

    //Population method generates a random initial population with
        its population costs
    private void population() {
      for(int i=0;i<n;i++){
        population.add(solution());
        popCost.add(JSSP.fitness(population.get(i)));
      }
    }

    //popCost contains the fitness value of each individial in the
        same order
    private static void popCost() {
```

```
    popCost.clear();
    for(int i=0;i<n;i++){
      popCost.add(JSSP.fitness(population.get(i)));
    }
}

//Generate a random solution
private ArrayList<Integer> solution(){
  ArrayList<Integer> solution=new ArrayList<>();
  int temp,rnd;
  for(int i=0;i<job*machine;i++)
    solution.add(i%job);
  for(int j=0;j<job*machine;j=j+4){
    temp=solution.get(j);
    rnd=(int)(Math.random()*job*machine);
    solution.set(j, solution.get(rnd));
    solution.set(rnd, temp);
  }
  return solution;
}

//Sort the population respecting to the individuals fitness value
private static void sortPopulation() {
  ArrayList<ArrayList<Integer>> newPop=new ArrayList();
  int i=0,j=0;
  newPop.add(j, DCSJSSP.cloner(population.get(i++)));
  do{
    boolean inc=false;
    while(!inc && j<newPop.size()){
      if(popCost.get(i)<JSSP.fitness(newPop.get(j))){
        newPop.add(j, DCSJSSP.cloner(population.get(i)));
        inc=true;
          }
      else j++;
      }
    if(!inc) newPop.add(j, DCSJSSP.cloner(population.get(i)));
    j=0;
    i++;
  }while(i<n);
  population=newPop;
  popCost();
}

//Search for new nests with the smart portion of cuckoos
private void cs() {
    ArrayList<Integer> sol;
    double levy;
  for(int i=1;i<n*pc;i++){
    levy=levy(mu);
    sol=DCSJSSP.cloner(population.get(i));
    if(levy<0.3){
      sol=DCSJSSP.localSearch(DCSJSSP.perturbSwap
          (DCSJSSP.perturbInsert(sol)));}
```

```
    else if(levy<0.6)
      sol=DCSJSSP.localSearch(DCSJSSP.perturbInsert
          (DCSJSSP.perturbInvers(sol)));
    else
      sol=DCSJSSP.localSearch(DCSJSSP.perturbInvers
          (DCSJSSP.perturbInvers(sol)));
    if(popCost.get(i)>JSSP.fitness(sol)){
      population.set(i, DCSJSSP.localSearch(sol));
        popCost.set(i, JSSP.fitness(population.get(i)));}
   }
  sortPopulation();
}


//Get a cuckoo and choose randomly an individual.
//The better the cuckoo and the individual takes a nest in the
    population
private void getCukoo() {
  ArrayList<Integer> sol=DCSJSSP.cloner(population.get(0));
  double levy=levy(mu);
  int k=(int) (Math.random()*n);
  if(levy<0.5)
    sol=DCSJSSP.localSearch(DCSJSSP.perturbSwap
        (DCSJSSP.perturbSwap(sol)));
  else
    sol=DCSJSSP.localSearch(DCSJSSP.perturbInsert
        (DCSJSSP.perturbInsert(sol)));
  sol=DCSJSSP.localSearch(sol);
  if(popCost.get(k)>JSSP.fitness(sol))
    {population.set(k, sol);}
   popCost();
   sortPopulation();
}


//Replace a portion p_a of worst individuals by new better
    cuckoos
private void worst() {
   ArrayList<Integer> sol;
   for(int i=n-(int)(n*pa);i<n;i++){
      sol=DCSJSSP.cloner(population.get(0));
      sol=DCSJSSP.localSearch(DCSJSSP.perturbSwap
          (DCSJSSP.perturbSwap(sol)));
   sol=DCSJSSP.localSearch(sol);
    if(popCost.get(i)>JSSP.fitness(sol))
      {population.set(i,sol);
      popCost.set(i, JSSP.fitness(sol));}
   }
   sortPopulation();
}


//A simple approximation of Levy flights distribution
    private static double levy(double mu){
```

```
        double t=Math.random()*9+1;//
          t=Math.pow((1/t),(mu));
        return t;
    }
}
```

B.3   DCS for QAP

B.3.1   QAP

Each QAP file is read and handled by this class to generate flow and distance matrices
(matrix 1 and matrix 2) that contains the flows and distances between each couple
of facility sites. Additionally, it returns the fitness value for a given solution.

```
package dcsqap;

import java.io.File;
import java.io.FileNotFoundException;
import java.util.ArrayList;
import java.util.Scanner;
import java.util.StringTokenizer;

public class QAP {
    //Instance dimension (number of facilities)
    public static int dimension;
    //Flow and distance matrices
    public static ArrayList<ArrayList<Integer>> matrix1=new ArrayList<>();
    public static ArrayList<ArrayList<Integer>> matrix2=new ArrayList<>();

    public QAP(File f){
        matrix(f);}

// Read file information
  public void matrix(File f){
    Scanner input1=null;
    try {
      input1=new Scanner(f);
    } catch (FileNotFoundException e) {
      System.out.println(e.toString());
    }StringTokenizer st;
    if(input1.hasNextLine()){
      st=new StringTokenizer(input1.nextLine()," ");
      dimension=Integer.valueOf(st.nextToken());
      }
    int i=0;
    while(input1.hasNextLine()&& i<=dimension){
      StringTokenizer str=new StringTokenizer(input1.nextLine()," ");
      ArrayList<Integer> l=new ArrayList<>();
      while(str.hasMoreTokens()){
        l.add(Integer.valueOf(str.nextToken()));
```

```
      }
    if(!l.isEmpty()){
       matrix1.add(l);
    }
    i++;
  }
  while(input1.hasNextLine()){
     StringTokenizer str=new StringTokenizer(input1.nextLine()," ");
     ArrayList<Integer> l=new ArrayList<>();
     while(str.hasMoreTokens()){
         l.add(Integer.valueOf(str.nextToken()));
     }
     if(!l.isEmpty()){
         matrix2.add(l);
     }
  }
}

//Fitness of a solution sol
public static int fitness(ArrayList<Integer> sol){
  int fit=0;
  for(int i=0;i<sol.size();i++)
    for(int j=0; j<sol.size(); j++){
       fit+=matrix1.get(i).get(j)*matrix2.get(sol.get(i)).get(sol.get(j));
       }
    return fit;
  }
}
```

B.3.2 DCSQAP

Each QAP solution is a series of assignments of facilities with sites. The solution is then represented by a series of integer numbers and adaptations of DCS to this type of instances, which is provided by DCSQAP class.

```
package dcsqap;
import java.io.File;
import java.io.IOException;
import java.util.ArrayList;

public class DCSQAP {

  //Contains the instance information
  static QAP qap;
   //Instance dimension
  static int dim;

  public static void main(String[] args) throws IOException{
     File f=new File("tai15a.dat");
     qap=new QAP(f); dim=QAP.dimension;
```

```
      DCS dcs=new DCS(0.2,0.6,2,40,500);
      //Run DCS
      dcs.run();
      //Print result
    System.out.println("best "+DCS.popCost.get(0));
  }


  //Local search around a solution ''sol''
  public static ArrayList<Integer> localSearch(ArrayList<Integer>
        sol) {
    ArrayList<Integer> sol1=cloner(sol);
    for(int i=0;i<dim-1;i++){
      for(int j=i+1;j<dim;j++){
        if(gain(sol1,i,j)<0){
          sol1=swap(sol1, i, j);
        }
      }
    }
      return sol1;
  }

  //Big step from a solution sol
  //It is performed by swapping 4 random facilities
  public static ArrayList<Integer> bigStep(ArrayList<Integer> sol) {
    ArrayList<Integer> sol1=sol;
    for(int i=0;i<4;i++){
      sol1=swap(sol1,i,(int)(Math.random()*(dim-1)));
      }
    return sol1;
  }

  //swap move on sol between facilities p1 and p2
  public static ArrayList<Integer> swap(ArrayList<Integer> sol, int
      p1, int p2) {
    ArrayList<Integer> sol1=cloner(sol);
    sol1.set(p1, sol.get(p2));
    sol1.set(p2, sol.get(p1));
    return sol1;
  }

  //small step on sol by applying a random swap
  public static ArrayList<Integer> smallStep(ArrayList<Integer>
      sol) {
    int p1=(int)(Math.random()*(dim-1));
    int p2=(int)(Math.random()*(dim-1));
    return swap(sol, p1 , p2);
  }

  //The gain of a swap move between facilities i and j in a
        solution sol
  public static int gain(ArrayList<Integer> sol, int i, int j){
    int d = (QAP.matrix1.get(i).get(i)-QAP.matrix1.get(j).get(j))*
```

```
    (QAP.matrix2.get(sol.get(j)).get(sol.get(j))
    -QAP.matrix2.get(sol.get(i)).get(sol.get(i))) +
    (QAP.matrix1.get(i).get(j)-QAP.matrix1.get(j).get(i))*
    (QAP.matrix2.get(sol.get(j)).get(sol.get(i))
    -QAP.matrix2.get(sol.get(i)).get(sol.get(j)));
    for (int k = 0; k < QAP.dimension; k++)
      if (k!=i && k!=j)
      d = d + (QAP.matrix1.get(k).get(i)-QAP.matrix1.get(k).get(j))*
      (QAP.matrix2.get(sol.get(k)).get(sol.get(j))
      -QAP.matrix2.get(sol.get(k)).get(sol.get(i))) +
      (QAP.matrix1.get(i).get(k)-QAP.matrix1.get(j).get(k))*
      (QAP.matrix2.get(sol.get(j)).get(sol.get(k))
      -QAP.matrix2.get(sol.get(i)).get(sol.get(k)));
    return d;
  }

  //Clone a solution assignment
  public static ArrayList<Integer> cloner(ArrayList<Integer>
      assignment) {
    ArrayList<Integer> fac=new ArrayList<>();
    for(int i=0;i<assignment.size();i++){
        fac.add(assignment.get(i));
    }
    return fac;
  }

}
```

B.3.3   DCS

This class contains the same structure as mentioned in the other DCS classes. They are reported separately to show the differences between them with respect to each set of problem instances.

```
package dcsqap;
import java.util.ArrayList;

public class DCS {
    public static ArrayList<ArrayList<Integer>> population=new
        ArrayList<>();
//Population solution costs (fitnesses)
    public static ArrayList<Integer> popCost=new ArrayList<>();
//Population size
    public static int n;
//Max generations
```

```java
public int maxgen;
public double pa;
public double pc;
public double mu;

 public int dim;

DCS(double a,double c, double m, int np, int max){
  pa=a; pc=c; mu=m; n=np; maxgen=max;
  dim=QAP.dimension;
  }

 public void run() {
   //Initialize the population
   population();
   //Sort the population: the best is in the first position
   sortPopulation();
   //Launch the algorithm
   while(--maxgen>0){
     cs();
     getCukoo();
     worst();
   }
  //Print the result
  System.out.println("result "+popCost.get(0));
  }


///////////////////////****************************/////////////////////////

//Population method generates a random initial population with its
   population costs
  private void population() {
    for(int i=0;i<n;i++){
     population.add(solution());
     popCost.add(QAP.fitness(population.get(i)));
    }
  }

 //popCost contains the fitness value of each individial in the
     same order
   private static void popCost() {
     popCost.clear();
     for(int i=0;i<n;i++){
       popCost.add(QAP.fitness(population.get(i)));
     }
   }
   //Generate a random solution
   private ArrayList<Integer> solution() {
     ArrayList<Integer> solution=new ArrayList<>();
     int temp,rnd;
     for(int i=0;i<dim;i++)
         solution.add(i);
     for(int j=0;j<dim;j=j+4){
```

```
      temp=solution.get(j);
      rnd=(int)(Math.random()*dim);
      solution.set(j, solution.get(rnd));
      solution.set(rnd, temp);
    }
    return solution;
}
//Sort the population respecting to the individuals fitness value
private static void sortPopulation() {
    ArrayList<ArrayList<Integer>> newPop=new ArrayList<>();
    int i=0,j=0;
    newPop.add(j, DCSQAP.cloner(population.get(i++)));
    do{
      boolean inc=false;
      while(!inc && j<newPop.size()){
        if(popCost.get(i)<QAP.fitness(newPop.get(j))){
          newPop.add(j, DCSQAP.cloner(population.get(i)));
          inc=true;
        }
        else j++;
      }
      if(!inc) newPop.add(j, DCSQAP.cloner(population.get(i)));
      j=0;
      i++;
    }while(i<n);
    population=newPop;
    popCost();
}

//Search for new nests with the smart portion of cuckoos
private void cs() {
  ArrayList<Integer> sol;
  double levy;
  for(int i=1;i<n*pc;i++){
      levy=levy(mu);
    sol=DCSQAP.cloner(population.get(i));
    if(levy<0.5){
      sol=DCSQAP.localSearch(DCSQAP.smallStep(sol));}
    else
      sol=DCSQAP.localSearch(DCSQAP.bigStep(sol));

    if(popCost.get(i)>QAP.fitness(sol))
      {population.set(i, DCSQAP.localSearch(sol));
        popCost.set(i, QAP.fitness(population.get(i)));}
    }
    sortPopulation();
}
```

```
//Get a cuckoo and choose randomly an individual.
  //The better the cuckoo and the individual takes a nest in the
       population
  private void getCukoo() {
    ArrayList<Integer> sol=DCSQAP.cloner(population.get(0));
    double levy=levy(mu);
    int k=(int) (Math.random()*(n-(n*pa)-(n*pc))+n*pa);
    if(levy<0.5){
      sol=DCSQAP.localSearch(DCSQAP.smallStep(sol));}
    else
      sol=DCSQAP.localSearch(DCSQAP.bigStep(sol));
    sol=DCSQAP.localSearch(sol);
    if(popCost.get(k)>QAP.fitness(sol))
      {population.set(k, sol);}
    popCost();
    sortPopulation();
  }

  //Replace a portion p_a of worst individuals by new better
       cuckoos
  private void worst() {
    ArrayList<Integer> sol;
    for(int i=(int)(n-(n*pa));i<n;i++){
      sol=DCSQAP.cloner(population.get(0));
      sol=DCSQAP.localSearch(DCSQAP.bigStep(sol));
      sol=DCSQAP.localSearch(sol);
      if(popCost.get(i)>QAP.fitness(sol))
        {population.set(i,sol);
           popCost.set(i, QAP.fitness(sol));}
    }
      sortPopulation();
  }

  //A simple approximation of Levy flights distribution
  private static double levy(double mu){
    double t=Math.random()*9+1;//
    t=Math.pow((1/t),(mu));
    return t;
  }
}
```

B.4   RKCS for TSP

B.4.1   TSP

```
package rkcstsp;
import java.io.File;
```

```java
import java.io.FileNotFoundException;
import java.util.ArrayList;
import java.util.LinkedList;
import java.util.Scanner;
import java.util.StringTokenizer;

public class Tsp {
  //Instance information
   private String nom;
   private String comment="";
   private String type;
   private String edge_weight_type;
   private String dimension;
    //Coordinates of the cities
   public static ArrayList<ArrayList<Double>> node_coord_section=new
       ArrayList<>();
    //Distance matrix
   public static ArrayList<ArrayList<Integer>> DisMatrix=new
       ArrayList<ArrayList<Integer>>();

    public Tsp(){}
    public Tsp(File f){
      tspName(f);
      coordonnee(f);
      DisMatrix=DistanceMatrix();
    }
//Reading the head information from the TSP instance file
  public void tspName(File f) {
    Scanner input = null;
    try {input = new Scanner(f);}
    catch (FileNotFoundException e) {
       System.out.println(e.toString());}
    String info;
    while(input.hasNextLine()){
      StringTokenizer str=new StringTokenizer(input.nextLine(),":");
      while(str.hasMoreTokens()){
        info=str.nextToken();
        if(info.contains("NAME")) nom=str.nextToken();
        if(info.contains("COMMENT")) comment+=str.nextToken()+"\n";
        if(info.equalsIgnoreCase("TYPE ")) type=str.nextToken();
        if(info.contains("DIMENSION")) dimension=str.nextToken();
        if(info.contains("EDGE_WEIGHT_TYPE"))
            edge_weight_type=str.nextToken();
      }
    }
  }
//Cities coordinates from 0 to n-1
  public void coordonnee(File f){
    Scanner input1=null;
    int dex=0;
    try {input1=new Scanner(f);}
     catch (FileNotFoundException e) {
      System.out.println(e.toString());}
    while(input1.hasNextLine()){
      StringTokenizer str1=new StringTokenizer(input1.nextLine()," ");
      String info=" ";
```

```
      if(str1.hasMoreTokens()) info=str1.nextToken();
      if(info.equalsIgnoreCase("EOF")||info.equalsIgnoreCase("-1")) dex=0;
      if(dex==1){
        ArrayList<Double> s=new ArrayList<>();
        StringTokenizer no1=new StringTokenizer(str1.nextToken()," ");
        s.add(Double.valueOf(no1.nextToken()));
        StringTokenizer no2=new StringTokenizer(str1.nextToken()," ");
        s.add(Double.valueOf(no2.nextToken()));
        node_coord_section.add(s);
      }
      if(info.equalsIgnoreCase("NODE_COORD_SECTION")) dex=1;
    }
  }
//Distance between two cities v1 and v2
  public double distance(ArrayList<Double> v1,ArrayList<Double> v2){
    return
        Math.sqrt(Math.pow((v1.get(0)-v2.get(0)),2)
        +Math.pow((v1.get(1)-v2.get(1)),2));
  }
//Distance matrix
    public ArrayList<ArrayList<Integer>> DistanceMatrix(){
      ArrayList<ArrayList<Integer>> matrix=new ArrayList<>();
      for(int i=0;i<node_coord_section.size();i++){
        ArrayList<Integer> ligne=new ArrayList<>();
        for(int j=0;j<node_coord_section.size();j++){
          ligne.add((int)(distance(node_coord_section.get(i),
            node_coord_section.get(j))+0.5));}
        matrix.add(ligne);
      }
      return matrix;
    }

    //Fitness function returns the length of sol cycle
    public static int fitness(LinkedList<City> sol){
      int len=0;
      int i;
      for(i=0;i<(nbrCities())-1;i++){
        len+=DisMatrix.get(sol.get(i).index).get(sol.get(i+1).index);}
      len+=DisMatrix.get(sol.get(i).index).get(sol.get(0).index);
      return len;
    }

    //Instance dimension
    public static int nbrCities(){
        return node_coord_section.size();
    }
}
```

B.4.2 CITY

The class CITY contains a city agent data like the position and the random-key value (used to be controlled by Lévy), and functions that control this agent. It contains also an approximation of Lévy flights distribution based on Lanczos Gamma approximation.

```java
package rkcstsp;
import java.util.Random;

public class City {
    // Random-key variable
    public double rnd;
    // City number
    public int index;
    // For generating random variables
    public static Random r1=new Random();

    public City(double r, int p){
        rnd=r;
        index=p;
    }

    public City(City c){
        rnd=c.rnd;
        index=c.index;
    }

    public City cloner(){
        return new City(this);
    }

    public double compareTo(City c){
        return rnd-c.rnd;
    }

    //Approximation of Levy flights
    public static double levy(double lambda){

        double
            t=(r1.nextGaussian()*sigma(lambda))/
                (Math.pow(Math.abs(r1.nextGaussian()),1/ lambda));
        return Math.abs(t*0.1)>=1?0.5:Math.abs(t*0.1);
    }

    //Approximation of the variance
    private static double sigma(double lambda) {
        return
```

```
        Math.pow((gamma(1+lambda)*Math.sin(Math.PI*lambda/2)/
          (gamma((1+lambda)/2)*lambda*Math.pow(2,((lambda-1)/2)))),
          (1/lambda));
    }

    //Approximation of Lanczos Gamma function
    static double logGamma(double x) {
     double tmp = (x - 0.5) * Math.log(x + 4.5) - (x + 4.5);
     double ser = 1.0 + 76.18009173 / (x + 0)  - 86.50532033   / (x + 1)
               + 24.01409822  / (x + 2)  -  1.231739516 / (x + 3)
               +  0.00120858003 / (x + 4) - 0.00000536382 / (x + 5);
     return tmp + Math.log(ser * Math.sqrt(2 * Math.PI));
    }
    static double gamma(double x) { return Math.exp(logGamma(x)); }

    //Perturb the current solution
    //Alpha must be out of the interval [rnd1,rnd2] of the precedent
        and the next cities
    public void perturb(double lambda, double rnd1, double rnd2){
       double bar=Math.random();
       int signe=bar>0.5?1:-1;
       double alpha=(bar>0.5?rnd2:rnd1)+levy(lambda)*signe;
       rnd= Math.abs(alpha)>1?2-Math.abs(alpha):Math.abs(alpha);
    }
}
```

B.4.3   CityList

CityList is the class that represents a solution and its initialization and perturbation functions.

```
package rkcstsp;
import java.util.ArrayList;
import java.util.LinkedList;
import java.util.ListIterator;

public class CityList extends LinkedList<City>{
   //Problem dimension
   final int dim=Tsp.nbrCities();;
    public CityList(){}

   public CityList(CityList cl){
     ListIterator it=cl.listIterator();
     while(it.hasNext()) add(new City((City)it.next()));}

   public CityList(ArrayList<Integer> list){
```

```
  for(int i=0; i<dim; i++)
    add(new City(0, list.get(i)));
}

// Initialization of a random solution
public void inisializ(){
  for(int i=0; i<dim; i++){
    City c=new City(Math.random(), i);
    // Insert the city c in its position among solution cities
    add(localiz(c),c);
  }
}

  // Localize the city position considering the value of rnd
  public int localiz(City c){
    int i=0;
    for(; i<size(); i++){
      if((c.rnd)<(this.get(i).rnd)) return i;
    }
    return i;
  }

  // Perturb the city in the position i
  public void perturb(int i, double lambda){
    get(i).perturb(lambda,get((i)<=0?size()-1:i-1).rnd,
      get((i+1)>=size()?0:i+1).rnd);
    City c=get(i).cloner();
    remove(i);
    int pos=localiz(c);
    add(pos,c);
  }

  // Perturb a random list of cities
  public CityList perturbList(double lambda){
    CityList cl=new CityList(this);
    // Random number
    double y=City.levy(lambda);
    // Random number of cities
    int x=2+Math.abs(((int)((dim-3)*y)));
    // A list of x cities
    ArrayList<Integer> list=nCity(x);
    // Perturb each city of the list
    for(int i=0; i<list.size(); i++){
    perturb(list.get(i),lambda);
    }
    return cl;
  }


  // Return a set of l cities chosen randomly
  public ArrayList<Integer> nCity(int l){
    ArrayList<Integer> list=new ArrayList<>();
    for(int i=0; i<l; i++)
```

```
        list.add(Math.abs((int)(Math.random()*dim)));
      return list;
    }

    //Duplicate the current solution
    public CityList cloner(){
      return new CityList(this);
    }

    public int getCost() {
      return Tsp.fitness(this);
    }

    //Replace the current solution by ''cuckoo''
    public void replace(CityList cuckoo) {
      clear();
      ListIterator it=cuckoo.listIterator();
      while(it.hasNext()) add(new City((City)it.next()));
    }
}
```

B.4.4   Population

Population is the class that contains the main adaptation functions of RKCS to TSP.

```
package rkcstsp;
import java.util.ArrayList;
import java.util.LinkedList;

public class Population extends LinkedList<CityList>{
  //Population size
    private final int size;
//Worst cuckoos' portion
    static double pa=0.2;
     //Costs of the population individuals
    public LinkedList<ArrayList<Integer>> costList=new
        LinkedList<>();

    public Population(int size){
      this.size=size;
    }

    //Initialize population
    public void initializ(){
      clear();
      for(int i=0; i<size; i++){
        CityList cl=new CityList();
```

```java
      cl.inisializ();
      insertCost(Tsp.fitness(cl),i);
      add(cl);
  }
}

// Placement of a city pos and its cost cl in the cost list
private void insertCost(int cl, int pos) {
  ArrayList<Integer> list=new ArrayList<>();
  list.add(cl);
  list.add(pos);
  int i;
  for(i=0; i<costList.size(); i++){
     if(list.get(0)<costList.get(i).get(0))
       break;
  }
  costList.add(i, list);
}

//Improving the individual m by searching around the best
    solution
 //If the found solution is better than m it will replace it
public void improvBest(double lambda, int m){
  CityList cl=get(costList.get(0).get(1)).perturbList(lambda);
  cl=localSearch(cl);
  if(cl.getCost()<get(m).getCost())
    replace(m,cl);
}

//Replacing worst cuckoo portion by new better ones
public void improvWorst(double lambda) {
  for(int i=(int)(size*(1-pa)); i<size; i++){
     improvBest(lambda, costList.get(i).get(1));
  }
}

//Improve Smart cuckoo solutions
public void improvCS(double lambda) {
  for(int i=0; i<(size*(1-pa)); i++)
     improvCS(lambda, costList.get(i).get(1));
}

//Smart cuckoo improvement
public void improvCS(double lambda,int m) {
  CityList cl=new CityList(get(m));
    cl.perturbList(lambda);
  cl=localSearch(cl);
  if(cl.getCost()<get(m).getCost())
      replace(m,cl);
}

//Get cuckoo function
public void getCuckoo(double lambda){
```

```
//Get a new cuckoo from the best position of the population
   CityList cuckoo=new CityList(get(costList.get(0).get(1)));
   cuckoo.perturbList(lambda);
   cuckoo=localSearch(cuckoo);
//Get the cuckoo fitness
  int quality=cuckoo.getCost();
//Get a random position in the population without worst
    individuals
  int pos=(int)(Math.random()*size*(1-pa));
//Replace the current cuckoo if the new one is better
  if(quality<costList.get(pos).get(0))
     replace(costList.get(pos).get(1),cuckoo);
}

//Local search around sol
public CityList localSearch(CityList sol) {
  int r=0;
  CityList sol1=new CityList(sol);
  for(int i=0;i<sol.dim-2;i++){
    if(i==0) r=0;
    for(int j=i+2;j<sol.dim-1+r;j++){
      if(gain(sol1,i,j)<0){
      sol1=smallStep(sol1, i, j);
    }
  }
 }
 r=1;
 }
 return sol1;
}

//Small step from sol by applying 2-opt move between p1 and p2
   cities
private static CityList smallStep(CityList sol, int p1, int p2) {
    CityList sol1=new CityList(sol);
    for(int i=p1+1,j=p2;i<j;i++,j--){
      sol1.get(i).index=(sol.get(j).index);
      sol1.get(j).index=(sol.get(i).index);
    }
    return sol1;
}

//Gain of 2-opt move from sol1
public int gain(CityList sol1, int i, int j) {
  int x=Tsp.DisMatrix.get(sol1.get(i).index).
      get(sol1.get(i+1).index);
  int y=Tsp.DisMatrix.get(sol1.get(j).index).
      get(sol1.get((j+1)%(sol1.dim)).index);
  int u=Tsp.DisMatrix.get(sol1.get(i).index).
      get(sol1.get(j).index);
  int v=Tsp.DisMatrix.get(sol1.get(i+1).index).
      get(sol1.get((j+1)%(sol1.dim)).index);
  int a=(u+v)-(x+y);
  return a;
```

```
    }

    //Put cuckoo in the nest ''nest''
    public void replace(int nest, CityList cuckoo) {
     get(nest).replace(cuckoo);
     updateCosts(nest,Tsp.fitness(cuckoo));
    }

    //Update the cost of the city in pos by the new cost
    public void updateCosts(int pos, int cost){
      for(int i=0; i<costList.size(); i++)
        if(costList.get(i).get(1)==pos)
          costList.remove(i);
      insertCost(cost, pos);
    }

}
```

B.4.5   RKCSTSP

RKCSTSP class represents the structure and steps to be followed by the main program with respect to CS algorithm (the improved version).

```
package rkcstsp;
import java.io.File;

public class RKCSTSP {
static String city="berlin52";
//Population size
   static int n=20;
//Creating an n size population
   static Population population=new Population(n);
//Maximum generation
   public static int MaxGen=50;
//Levy parameter
   static double lambda=3/2;

  public static void main(String[] args) {
     getTsp(city);
//initial population
   population.initializ();
//Stopping criteria
     while(--MaxGen>0){
         population.getCuckoo(lambda);
//Smart cuckoos phase
     population.improvCS(lambda);
//Replace worst individuals by better new ones
       population.improvWorst(lambda);
     }
```

```
        //Print the results
        System.out.println(population.costList.get(0).get(0));
    }
//Read the problem instance and create an object containing its data
    public static void getTsp(String instance){
        File f=new File(instance+".tsp");
        new Tsp(f);
    }


}
```

B.5   RKCS for QAP

B.5.1   QAP

```
package rkcsqap;
import java.io.File;
import java.io.FileNotFoundException;
import java.util.ArrayList;
import java.util.LinkedList;
import java.util.Scanner;
import java.util.StringTokenizer;

public class Qap {
  //Instance information
   public static int dimension;
    //Flow and distance matrices
   public static ArrayList<ArrayList<Integer>> matrix1=new ArrayList<>();
   public static ArrayList<ArrayList<Integer>> matrix2=new ArrayList<>();

   public Qap(File f){
     matrix(f);}

//Reading the head information from the QAP instance file
   public void matrix(File f){
     Scanner input1=null;
     try {
       input1=new Scanner(f);
     } catch (FileNotFoundException e) {
       System.out.println(e.toString());
     }StringTokenizer st;
     if(input1.hasNextLine()){
       st=new StringTokenizer(input1.nextLine()," ");
       dimension=Integer.valueOf(st.nextToken());
       }
     int i=0;
```

```
   while(input1.hasNextLine()&& i<=dimension){
     StringTokenizer str=new StringTokenizer(input1.nextLine()," ");
       ArrayList<Integer> l=new ArrayList<>();
     while(str.hasMoreTokens()){
       l.add(Integer.valueOf(str.nextToken()));
     }
     if(!l.isEmpty()){
         matrix1.add(l);
     }
     i++;
   }
   while(input1.hasNextLine()){
     StringTokenizer str=new StringTokenizer(input1.nextLine()," ");
       ArrayList<Integer> l=new ArrayList<>();
     while(str.hasMoreTokens()){
       l.add(Integer.valueOf(str.nextToken()));
     }
     if(!l.isEmpty()){
       matrix2.add(l);
     }
   }
 }

 //Fitness of the solution fac
 public static int fitness(LinkedList<Facility> fac){
   int cost=0;
   for(int i=0;i<fac.size();i++)
     for(int j=0; j<fac.size(); j++){
     cost+=matrix1.get(i).get(j)*matrix2.get(fac.get(i).pos).
           get(fac.get(j).pos);
   }
     return cost;
 }
}
```

B.5.2   Facility

The class Facility controls the facility agent in the same way as the City class. It contains facility agent date (position and random key) and functions that control its move via Lévy flights.

```
package rkcsqap;
import java.util.Random;

public class Facility {
   // Random-key variable
   public double rnd;
```

```
// Facility number
public int pos;
// For generating random variables
public static Random rl=new Random();

public Facility(double r, int p){
  rnd=r;
  pos=p;
}

public Facility(Facility c){
  rnd=c.rnd;
  pos=c.pos;
}

public Facility cloner(){
  return new Facility(this);
}

public double compareTo(Facility c){
  return rnd-c.rnd;
}

//Approximation of Levy flights
public static double levy(double lambda){
  double t=(rl.nextGaussian()*sigma(lambda))/
   (Math.pow(Math.abs(rl.nextGaussian()),1/ lambda));
  return Math.abs(t*0.1)>=1?0.5:Math.abs(t*0.1);
 }

//Approximation of the variance
private static double sigma(double lambda) {
  return Math.pow((gamma(1+lambda)*Math.sin(Math.PI*lambda/2)/
         (gamma(((1+lambda)/2)*lambda*Math.pow(2,((lambda-1)/2)))),
         (1/lambda));
}

//Approximation of Lanczos Gamma function
static double logGamma(double x) {
  double tmp = (x - 0.5) * Math.log(x + 4.5) - (x + 4.5);
  double ser = 1.0 + 76.18009173 / (x + 0) - 86.50532033  / (x + 1)
             + 24.01409822  / (x + 2)  -  1.231739516 / (x + 3)
             +  0.00120858003 / (x + 4) - 0.00000536382 / (x + 5);
  return tmp + Math.log(ser * Math.sqrt(2 * Math.PI));
}
static double gamma(double x) { return Math.exp(logGamma(x)); }

 //Perturb the current solution
//Alpha must be out of the interval [rnd1,rnd2] of the precedent
    and the next facilities
 public void perturb(double lambda, double x1, double x2){
   double bar=Math.random();
   int signe=bar>0.5?1:-1;
   double alpha=(bar>0.5?x2:x1)+levy(lambda)*signe;
```

```
      rnd= Math.abs(alpha)>1?2-Math.abs(alpha):Math.abs(alpha);
  }
}
```

B.5.3   Facility list

A QAP solution is represented by a linked list that contains in each element a Facility object. By the reported functions, the class FacilityList initializes, clones, and perturbs the QAP solution.

```
package rkcsqap;
import java.util.ArrayList;
import java.util.LinkedList;
import java.util.ListIterator;


public class FacilityList extends LinkedList<Facility>{

  public FacilityList(){
  }

    public FacilityList(FacilityList cl){
      ListIterator it=cl.listIterator();
      while(it.hasNext()) add(new Facility((Facility)it.next()));
    }

    public FacilityList(ArrayList<Integer> list){
      for(int i=0; i<list.size(); i++)
        add(new Facility(0, list.get(i)));
    }

    // Initialization of an individual
    public void inisializ(){
      add(new Facility(Math.random(),0));
      for(int i=1; i<Qap.dimension; i++){
        Facility c=new Facility(Math.random(), i);
        // Insertion of the new individual in its position regarding
          its order
        add(localiz(c),c);
      }
    }

    // Localize the facility position considering the value of rnd
    public int localiz(Facility c){
      int i=0;
      for(; i<size(); i++){
        if(c.compareTo((Facility)this.get(i))<0) return i;
      }
      return i;
```

```java
  }


  // Perturb the facility in the position i
  public void perturb(int i, double lambda){
    get(i).perturb(lambda,get((i)<=0?size()-1:i-1).rnd,
      get((i+1)>=size()?0:i+1).rnd);
    Facility c=get(i).cloner();
    remove(i);
    add(localiz(c),c);
  }


  // Perturb a random list of facilities
  public void perturbRndList(double lambda){
    // Random number
    double y=Facility.levy(lambda);
    // Random number of facilities
    int x=2+Math.abs(((int)((Qap.dimension-3)*y)));
    // A list of x facilities
    ArrayList<Integer> list=nFacility(x);
    // Perturb each facility of the list
    for(int i=0; i<x; i++)
      perturb(list.get(i),lambda);
  }


  // Return a set of l facilities chosen randomly
  public ArrayList<Integer> nFacility(int l){
    ArrayList<Integer> list=new ArrayList<>();
    for(int i=0; i<l; i++)
        list.add(Math.abs((int)(Math.random()*Qap.dimension)));
      return list;
  }


  //Duplicate the current solution
  public FacilityList clone(){
    return new FacilityList(this);
  }

  public int getCost() {
    return Qap.fitness(this);
  }

  //Replace the current solution by ``cuckoo''
  public void replace(FacilityList cuckoo) {
    clear();
    ListIterator it=cuckoo.listIterator();
    while(it.hasNext()) add(new Facility((Facility)it.next()));
  }
}
```

B.5.4   Population

The main functions that are used for adapting CS to QAP instances are shown in Population class.

```java
package rkcsqap;
import java.util.ArrayList;
import java.util.LinkedList;


public class Population extends LinkedList<FacilityList>{

  //Population size
  private final int size;
   //Costs of the population individuals
  public LinkedList<ArrayList<Integer>> costList=new LinkedList<>();

  public Population(int size){
    this.size=size;
  }

  //Initialize population
  public void initializ(){
    clear();
    for(int i=0; i<size; i++){
      FacilityList cl=new FacilityList();
      cl.inisializ();
      insertCost(Qap.fitness(cl),i);
      add(cl);
        }
  }

  // Placement of a city pos and its cost cl in the cost list
  private void insertCost(int cl, int pos) {
    ArrayList<Integer> list=new ArrayList<>();
    list.add(cl);
    list.add(pos);
    int i;
    for(i=0; i<costList.size(); i++){
      if(list.get(0)<costList.get(i).get(0))
        break;
    }
    costList.add(i, list);
  }

  //Improving the individual m by searching around the best solution
  //If the found solution is better than m it will replace it
  public void improvBest(double lambda, int m){
    FacilityList cl=cloner(get(costList.get(0).get(1)));
    cl.perturbRndList(lambda);
    cl=localSearch(cl);
    if(cl.getCost()<get(m).getCost())
      replace(m,cl);
  }

  //Replacing worst cuckoo portion by new better ones
  public void improvWorst(double lambda, double wc) {
    for(int i=(int)(size*(1-wc)); i<size; i++){
```

```
      improvBest(lambda, costList.get(i).get(1));
   }
}

//Improve Smart cuckoo solutions
public void improvCS(double lambda, double wc) {
  for(int i=0; i<(size*(1-wc)); i++)
     improvCS(lambda, costList.get(i).get(1));
}

//Smart cuckoo improvement
public void improvCS(double lambda,int m) {
  FacilityList cl=new FacilityList(get(m));
  cl.perturbRndList(lambda);
  cl=localSearch(cl);
  if(cl.getCost()<get(m).getCost())
     replace(m,cl);
}

//Get cuckoo function
public FacilityList getCuckoo(double lambda){
   FacilityList cuckoo=cloner(get(costList.get(0).get(1)));
  cuckoo.perturbRndList(lambda);
  return localSearch(cuckoo);
}

//Local search around sol
public FacilityList localSearch(FacilityList sol) {
  FacilityList sol1=new FacilityList(sol);
  for(int i=0;i<Qap.dimension;i++){
    for(int j=0;j<Qap.dimension;j++){
      if(gain(sol1,i,j)<0){
        sol1=swap(sol1, i, j);
      }
    }
  }
  return sol1;
}

//Small step from sol by applying swap move between p1 and p2 cities
private static FacilityList swap(FacilityList list, int p1, int p2) {
  FacilityList sol=cloner(list);
  sol.get(p1).pos=(list.get(p2).pos);
  sol.get(p2).pos=(list.get(p1).pos);
  return sol;
}

private static FacilityList cloner(FacilityList list) {
  return new FacilityList(list);
}
```

```
//Gain of swap move from sol1
private int gain(FacilityList sol1, int i, int j){
  int d;
  d = (Qap.matrix1.get(i).get(i)-Qap.matrix1.get(j).get(j))*
        (Qap.matrix2.get(sol1.get(j).pos).get(sol1.get(j).pos))
        -Qap.matrix2.get(sol1.get(i).pos).get(sol1.get(i).pos)
  +(Qap.matrix1.get(i).get(j)-Qap.matrix1.get(j).get(i))
          *(Qap.matrix2.get(sol1.get(j).pos).get(sol1.get(i).pos)
          -Qap.matrix2.get(sol1.get(i).pos).get(sol1.get(j).pos));
  for (int k = 0; k < Qap.dimension; k++)
    if (k!=i && k!=j)
      d = d + (Qap.matrix1.get(k).get(i)-Qap.matrix1.get(k).get(j))*
      (Qap.matrix2.get(sol1.get(k).pos).get(sol1.get(j).pos)
            -Qap.matrix2.get(sol1.get(k).pos).get(sol1.get(i).pos)) +
      (Qap.matrix1.get(i).get(k)-Qap.matrix1.get(j).get(k))*
      (Qap.matrix2.get(sol1.get(j).pos).get(sol1.get(k).pos)
            -Qap.matrix2.get(sol1.get(i).pos).get(sol1.get(k).pos));
  return d;
}

//Put cuckoo in the nest ''nest''
public void replace(int nest, FacilityList cuckoo) {
  get(nest).replace(cuckoo);
  updateCosts(nest,Qap.fitness(cuckoo));
}

//Update the cost of the city in pos by the new cost
public void updateCosts(int pos, int cost){
  for(int i=0; i<costList.size(); i++)
    if(costList.get(i).get(1)==pos)
            costList.remove(i);
    insertCost(cost, pos);
}
}
```

B.5.5 RKCSQAP

RKCSQAP is the class that contains the algorithm parameters and the main function presenting the structure of the program.

```
package rkcsqap;
import java.io.File;
import java.io.IOException;


public class CSRKQAP {
//Population size
   static int n=30;
//Creating an n size population
   static Population population=new Population(n);
//Maximum generation
   public static int MaxGen=500;
//Levy parameter
   static double lambda=3/2;

   static double pa=0.2;
```

```
    public static void main(String[] args) throws IOException{
        int best;
        getQap();
//initial population
        population.initializ();

//Stopping criteria
        while(--MaxGen>0){
          FacilityList cuckoo = population.getCuckoo(lambda);
          int quality=cuckoo.getCost();
          int pos=(int)(Math.random()*n*(1-pa));
          if(quality<population.costList.get(pos).get(0))
              population.replace(population.costList.get(pos).get(1),cuckoo);
//Smart cuckoos phase
        population.improvCS(lambda,pa);
//Replace worst individuals by better new ones
        population.improvWorst(lambda, pa);
    }
    best=population.costList.get(0).get(0);
     //Print the results
    System.out.println("best "+best);
    }

//Read the problem instance and create an object containing its data
    public static void getQap(){
        File f=new File("tai15a.dat");
        new Qap(f);
    }
}
```

Printed in the United States
by Baker & Taylor Publisher Services

Printed in the United States
by Baker & Taylor Publisher Services